Philosophy for Public Health and Public Policy

James Wilson is Professor of Philosophy, University College London. His research uses philosophy to address real world problems, including how these problems reveal gaps and weaknesses in existing philosophical theories. He has published widely on public health ethics, and also on the ownership and governance of ideas and information. He provides ethics advice to a range of public sector organisations, and is a member of the National Data Guardian's Panel.

Philosophy for Public Health and Public Policy

Beyond the Neglectful State

JAMES WILSON

OXFORD
UNIVERSITY PRESS

OXFORD
UNIVERSITY PRESS

Great Clarendon Street, Oxford, OX2 6DP,
United Kingdom

Oxford University Press is a department of the University of Oxford.
It furthers the University's objective of excellence in research, scholarship,
and education by publishing worldwide. Oxford is a registered trade mark of
Oxford University Press in the UK and in certain other countries

First published 2021
First published in paperback 2023

Published in the United States of America by Oxford University Press
198 Madison Avenue, New York, NY 10016, United States of America

British Library Cataloguing in Publication Data
Data available

Library of Congress Cataloging in Publication Data
Data available

ISBN 978-0-19-284405-7 (Hbk.)
ISBN 978-0-19-890058-0 (Pbk.)

Contents

PART II. BEYOND THE NEGLECTFUL STATE: AN ETHICAL FRAMEWORK FOR PUBLIC HEALTH

PART III. STRUCTURAL JUSTICE

Preface

This book aims to help you think more clearly about the role of the state in protecting and promoting health. Doing this requires reflection on a number of deep questions about public policymaking, the nature of values and how to think rigorously about them, what health is, and why protecting and promoting health should be a commitment role for any state worth living in.

Many people, even health professionals, have a view of health, and the measures that states should take to protect it, which is *biomedical* and *reactive*. Biomedical models take medicine to be a science that aims to prevent disease and repair bodily malfunctioning, and health to be simply the absence of disease. Reactive health systems wait for individuals to fall ill and then aim to restore them (in so far as it is possible) to full health, rather than focusing their energies on preventing people from becoming ill in the first place. A reactive biomedical approach is very likely to end up systematically prioritizing expensive hi-tech last-ditch treatments above less dramatic long-term measures that could save more lives more cheaply over the longer term, even as its proponents claim to take prevention seriously.

By the end of this book, I aim to persuade you to shift to a richer and more proactive way of thinking about health, and the state's role in protecting and promoting it. This richer approach draws on the idea of the human right to health, and in particular the idea that this right entails a right to public health. Doing so requires that we look beyond debates about whether there should be a right to health*care* to the broader domain of a right to the conditions necessary for healthy life.

Placing the protection and promotion of health at the heart of government policy has often proved not only to be difficult to achieve in practice, but also ethically controversial, with some accusing such policies of being 'Nanny State' or 'healthism' or of illicit interference with the liberties of individuals or even of corporations. I argue that this is to get things almost entirely backwards. Safety and security are the bedrock of a good life, and the Neglectful State is at least as terrifying a prospect as the Nanny State. Neglectful States fail to pursue cost-effective and proportionate measures that would make life safer for all, and in so doing show a callous disregard for the lives and well-being of everyone, but especially the vulnerable.

This is also a book about the nature of philosophy, and what philosophy when done well can contribute to public policy. I argue that philosophy is not just useful, but vital, for thinking coherently about not only health policy, but public policy more broadly. But in order for philosophy to play the role that it needs to play, philosophers themselves need to become much more comfortable with the contingent, the messy, and the political—an approach that draws on the pragmatist tradition in philosophy.

Acknowledgements

I first got interested in public health ethics through conversations with Angus Dawson in around 2007, when we were both at the Centre for Professional Ethics at Keele University. Angus has continued to be a guiding influence ever since. University College London, where I have been based since 2008, has provided an ideal environment to undertake the research that led to this book. For a number of years from 2009 onwards, I taught jointly with Jonathan Wolff on topics in health justice at a national and global level. Along with other members of the then Centre for Philosophy, Justice and Health, we convened a number of workshops and conferences on justice and health equity.

From 2012, I became convener of the UCL's MA Philosophy, Politics and Economics of Health. Most of the material in this book has been discussed with multiple cohorts of students from the course, and I am grateful for their insights and challenges. I have also been lucky to work with a group of very gifted research students—some of whose work I respond to in the book, including Gabriele Badano, Ben Fardell, Jasper Littmann, Polly Mitchell, Alexandre Gajevic Sayegh, and Francisca Stutzin. Particularly in the book's discussions of prioritization, I have benefited from a long association with the multidisciplinary UCL-KCL Social Values Group: Victoria Charlton, Clare Coultas, Katharina Kieslich, Peter Littlejohns, Catherine Max, Polly Mitchell, Annette Rid, Benedict Rumbold, Albert Weale, Gry Wester, and our other less frequent collaborators.

I have been helped greatly by a number of conversations over the years with Alan Cribb about the relationship between philosophy and social science, and in particular how to retain the trademark rigour and clarity of analytic philosophy, while reconstructing philosophy in a way that is far more contextual, dynamic, and aware of power relations. Alan's (2005) book *Health and the Good Society* (Oxford: OUP), was important for me as I first came to the topics, and I have continued to benefit from his advice and friendship ever since.

My approach to the philosophy of public policy has been shaped significantly by the time that I have spent working with policy professionals, and as a member of various national committees. A secondment to the Royal Society's Science Policy Centre as a Senior Policy Adviser in 2011–12, where I worked closely with Jessica Bland, Tony McBride, and Jack Stilgoe was particularly influential. From 2013 onwards I have been a member of a variety of national committees that advise the NHS in England on the ethics and governance data for planning, research, and commercial activity. From 2016 onwards, I have been a member of the National Data Guardian's panel of advisers, which has allowed me the opportunity to observe and contribute to policymaking at a national level. Those from whom I learned most include Joanne Bailey, the late Dame Fiona Caldicott, Alan Hassey, Sir Jonathan Montgomery, and Eve Sariyiannidou.

Various colleagues have commented on the manuscript in whole or in part. I am grateful in particular to Jonathan Anomaly, Richard Cookson, Alan Cribb, Angus Dawson, Vikki Entwhistle, Daniel Herron, Stephen John, Bruce Laurence, Polly Mitchell, Parashkev Nachev, Benedict Rumbold, Elizabeth Russell, Marcel Verweij, Verina Wild, and Jonathan Wolff.

Earlier versions of material in the book have been presented at the University of Antwerp, the Aristotelian Society, Queen's University Belfast, University of Birmingham, University of Bristol, the Brocher Foundation, Cardiff University, University of Cambridge's Centre for Science and Policy, Ghent University, Keele University, King's College London, University of Lancaster, University of London, École Normale Supérieure de Lyon, University of Manchester, University of Münster, the Nuffield Council on Bioethics, University of Oxford, the Pontifical Academy of Science, Vatican City, University of Roehampton, Ruhr University Bochum, Zurich University, and World Congresses of Bioethics in Rijeka, Singapore, Rotterdam, Mexico City, Edinburgh and Bengaluru. I thank all these audiences for their questions and comments.

Lastly, I should acknowledge the vital role of friends and family, with whom I have discussed these questions over a period of years—especially Victor Dura-Vila, Milena Nuti, and Raj Sehgal. To Abi Patrick I owe much more.

Chapter 2 incorporates some paragraphs initially published in James Wilson, 'Embracing Complexity: Theory, Cases and the Future of Bioethics', *Monash Bioethics Review*, 32 (2014): 3–21. Chapter 3.2 and 3.3, including Table 3.1, draws on 'Embracing Complexity'. Chapter 3.4–3.8 reworks James Wilson, 'Internal and External Validity in Thought Experiments', *Proceedings of the Aristotelian Society* 116(2) (2016): 1–26. Chapters 2.2 and 3.2 draw on material I initially wrote for an unpublished book manuscript on the philosophy and ethics translational research—thanks to my co-authors on that book, Sarah Edwards and Geraint Rees. Chapter 5 reworks James Wilson, 'Why It's Time to Stop Worrying About Paternalism in Health Policy', *Public Health Ethics* 4(3) (2011): 269–79, and some material from James Wilson, 'The Right to Public Health', *Journal of Medical Ethics* 42 (2016): 367–75. Chapter 6 draws heavily on James Wilson, 'The Right to Public Health'. Chapter 7.6 reworks a few paragraphs from James Wilson, 'Paying for Patented Drugs Is Hard to Justify: An Argument About Time Discounting and Medical Need', *Journal of Applied Philosophy* 29(3) (2012): 186–99.

Chapter 8.6 reuses a couple of paragraphs from James Wilson, 'Philanthro-capitalism and Global Health', in Solomon Benatar and Gillian Brock (eds), *Global Health: Ethical Challenges* (Cambridge: CUP, 2020), pp. 416–28. These paragraphs are reprinted with permission of Cambridge University Press. A few paragraphs of James Wilson, 'Health Inequities', in A. Dawson (ed.), *Public Health Ethics: Key*

Concepts in Policy and Practice (Cambridge: CUP, 2011) appear in a revised form in Chapter 9. These paragraphs are reprinted with permission of Cambridge University Press. An earlier version of a few paragraphs from Chapter 10.3, and one paragraph of Chapter 1, appeared in James Wilson, 'Who Owns the Effectiveness of Antibiotics?', in J. Coggon and S. Gola (eds), *Global Health and International Community: Ethical, Political and Regulatory Challenges* (London: Bloomsbury Academic, 2013), pp. 151–64. These paragraphs are reprinted with the permission of Bloomsbury Press. A few paragraphs of Chapter 10.4 draw on work I published with Tracey Chantler and Emilie Karafillakis, 'Vaccination—Is There a Place for Penalties for Non-Compliance?', *Applied Health Economics and Health Policy* 17(3) (2019): 265–71. Chapter 10.5 reworks material originally published in James Wilson, 'The Ethics of Disease Eradication', *Vaccine* 32 (2014): 7179–83.

1

Introduction

The steps that governments take, or fail to take, to protect our health make a significant difference to the lives we are able to lead. A longer life, and especially a longer healthier life, is thought by nearly all to be a benefit; but it is also a benefit that can sometimes be bought at too high a cost. Counteracting health risks may require not only significant financial outlay, but also that the state take a very active role in regulation, and interfere with citizens' choices in ways some find objectionable.

Putting health first would require a wholesale restructuring of priorities for most governments—a shift that some groups would welcome, but others vociferously oppose. Decision-makers the world over struggle to find a balance. It is common for public health policies to be both more interventionist than a sizeable minority of citizens are comfortable with, and yet still insufficient to bring down the incidence of particular diseases, especially where the drivers of increased health risk are contested or not easily within the control of government.

The rise of obesity provides a powerful, though unfortunately all-too-common example. The prevalence of obesity doubled in 73 of 195 countries from 1980 to 2015, and showed a sustained increase in most other countries (Afshin etxal. 2017). The rise of obesity matters because it is an important risk factor for conditions such as type 2 diabetes, coronary heart disease, stroke, osteoarthritis, and some cancers, among many others. The authors of the Global Burden of Disease study estimate that high body mass index (BMI) contributed to 4 million deaths per year in 2015, and 120 million years of healthy life lost globally.

While the prevalence of obesity has been rising nearly everywhere, prevalence also varies considerably from country to country: in 2017–18, prevalence of obesity among men was 42.7 per cent in the USA, and 27.4 per cent in England, while in Japan it was only 3.8 per cent (World Obesity Federation 2020). The combination of rising rates of obesity within countries, and significant differences between countries, invites analysis of the reasons for this. Despite the consensus that socially controllable factors make very significant differences to obesity rates, it is disputed precisely which factors are most relevant. A non-exhaustive list of proposed causes includes manual labour has been replaced by sedentary jobs; people are using cars more and other modes of transport such as walking or cycling less; fewer traditional home-cooked meals are being eaten and more prepackaged and processed food; fresh fruit and vegetables have become more expensive relative to processed foods, leading to 'food deserts';[1] food high in fat and sugar is cheap and widely available; portion sizes have increased; obesity has become normalized, with the size of others playing an anchoring role in perceptions of what is a healthy body image; and heavy use of antibiotics has shaped gut flora in ways that makes obesity more likely.

Overall, the picture is one of complex and contested explanations of obesity's rise, and a growing realization that even having a good understanding of the causes of this rise would not necessarily imply being able to prevent it (Kelly and Russo 2018). There is also wider disagreement about whether obesity reduction is even something with which the state should be involved (Anomaly 2012); and further questions about whether some potentially effective policies such as stigmatization of obesity are so ethically problematic as to rule them out from consideration (Callahan 2013; Ulijaszek 2017). Obesity presents just one of several cases in which policymakers face rising health problems without having an adequate understanding either of the causes of the increased incidence, or of workable and socially acceptable solutions.

The aim of this book is to use the tools of philosophy to shed light on both the ethical justification for state activity to protect and promote health, and the high-level strategies that policymakers should use to improve population health. Doing so will take us on a journey that illuminates a number of important questions about the nature of evidence in public policy, why policymakers need to adopt complex systems approaches to their work, and why philosophy itself needs to adapt and become more historical and contextual. I'll introduce the overall argument of the book in more detail towards the end of this introduction, but before we get there I need to say a few things about the role of ethical values in political decision-making, about health, and about public health in particular.

1.1 Ethical Values and Deliberative Communities

All public policy choices rely at crucial points on evaluative judgements and a core of ethical analysis and commitment. That is to say, they rely on judgements about what is good or bad, or right and wrong. Public health is no exception. Sometimes the ethical dimension of a policy choice is obvious, as when a country is deciding whether to legalize euthanasia or adjust its voting age. Even in cases that seem routine, policymakers must choose whether to pursue one path rather than another, and which programmes should be prioritized within the confines of their limited budget. These choices presuppose an evaluation of the policies in terms of the ethical values they will realize or instantiate.

---

[1] Food deserts are 'areas of relative exclusion where people experience physical and economic barriers to accessing healthy foods' (Reisig and Hobbiss 2000).

Sometimes attempts are made to disguise this fact. Choices are described and presented as if they were purely technical or financial, or as if there were no alternative. For any decision that matters, such descriptions are evasive or obfuscatory. Although cost–benefit analyses are sometimes presented as purely technical, they are suffused with ethical choices, from their understanding of what counts as a benefit and what counts as a cost, to how to measure these, and how to determine how the costs and benefits should be distributed, and whether (and if so, how) those who lose out should be compensated. Deciding that an intervention that would reduce risks of death is *too expensive* is already to be implicitly committed to certain ethical claims about the value of saving a life.

The values we live by and those we choose to embed into our institutions make significant practical differences.[2] Decisions about how and which values should inform policy are the most serious we face as political communities. The pursuit of 'right' answers to such questions is fraught with difficulty, as there is widespread disagreement not only about the values that should animate political decisions, but also about the way in which these values should be traded off or reconciled with one another.

What counts as a better answer to a value question, and who should decide is not obvious; but it is important not to fall into the trap of assuming that because these questions are difficult and controversial to answer, it is not worth taking the attempt to do so seriously. The sense that something deep is at stake, and that these choices are grave ones makes sense only against the background conviction that there are better and worse answers to these evaluative questions.

It seems unsatisfactory for several reasons to assume that politicians know best about how ethical values should be combined and traded off against one another. In a democracy, those who make the final decision on which politicians are elected are the people themselves. If politicians' expertise in values is crucial for the well-being of the state, then this would seem to call into question the rationale for having a democratic system in which all adult citizens, regardless of their own ethical expertise, get to choose which politicians should prevail. It would thus be an odd kind of democracy that took its elected politicians to be experts in values.

These questions about politicians' ethical expertise exist alongside, and are amplified by, their conflicts of interest. Conflicts of interest, where they exist, are apt to affect both evidence-seeking behaviour and judgements about value, meaning that those who experience conflicts of interest are likely to be biased in their judgements about what is in the public interest. Alongside other more specific interests that will vary on a case by case basis, politicians have interests in

---

[2] I understand values as guides or measures that can be used to aid reasoned decision-making. Ethical value is one type of value, but there is also, for example, financial or aesthetic value. In this book when I talk about values I will, unless otherwise stated, be talking about ethical values. Ethics, in the most general sense, is the academic discipline that aims at investigating what is right and wrong, and what is good for individuals and society.

being re-elected, as well as in their broader political party continuing to prosper. Though politicians may have difficulty in admitting this publicly (or even privately, if they have fully internalized their party lines), such interests are liable to conflict with the public interest—for example by adopting quick fixes or avoiding taking necessary, but unpopular, decisions. As Dewey put it, 'It is difficult for a person in a place of authoritative power to avoid supposing that what he wants is right as long as he has power to enforce his demand. And even with the best will in the world, he is likely to be isolated from the real needs of others, and the perils of ignorance are added to those of selfishness. History reveals the tendency to confusion of private privilege with official status' (Dewey and Tufts 1981: 226).[3]

Citizens disagree amongst themselves at least as much as politicians do. Citizenship and voting rights are usually open to everyone who is born within a territory, regardless of any test of ethical character or level of ethical competence. So while some citizens will have relevant expertise, it seems unlikely that this could be the case for all. And to the extent that there is relevant ethical expertise, majority voting would seem to be an ineffective way of mobilizing it, as there seems to be no guarantee that the ethically best choices and values will be the most popular.[4]

Citizens, too, face conflicts of interest between what would benefit them or those they identify with as private individuals, and what would be in the public interest. Business people want to reduce taxes and regulations. Opera lovers campaign to protect public subsidies for the arts. Even if it is agreed at a general level that a new toxic waste dump needs to be built, it is rare that a community will agree that theirs is the best place to site it.

Given the apparent difficulty in identifying and mobilizing ethical expertise, and the likelihood that any existing ethical expertise will in practice be overpowered by sectional interests, it might be tempting to give way to a pessimism about the role of values in politics, coming to accept that even if there is such a thing as expertise in ethical reasoning, it will play only a marginal role in politics.

My conviction is that we can do better: we must find ways of bringing not just values, but sound reasoning about values into the heart of policymaking. Doing this successfully is not easy, for both political and philosophical reasons.

---

[3] It might be argued that in a representative democracy, citizens do nonetheless trust politicians to take ethical decisions on their behalf, and they are entitled to place their trust in politicians if they want to. In response, it is important to distinguish between trust as an attitude, and trustworthiness. Trust can be placed either well or badly. Trust in someone who is not trustworthy is not a good thing (O'Neill 2002b). So, if citizens trust politicians to make ethical decisions on their behalf, the core question is whether they are wise to do so—and it is at this point that politicians' combination of lack of obvious expertise in ethical deliberation and conflicts of interest become salient once more.

[4] Partly as a result of this problem, deliberative approaches to democracy emphasize either an open textured account of deliberation that allows all to contribute on the basis of their particular and specific interests and expertise, or embody an autonomy-based approach that focuses on a right to be involved in decision-making that is not dependent on one's expertise in so doing.

Politically, the conflicts of interest that affect the judgements of both citizens and politicians will hinder attempts to construct a politics that reaches beyond narrow sectional interests. There are also a number of important difficulties in translating philosophical theories about what is valuable into actionable insights that will help guide policy on the ground. As Part I explores, for philosophers to provide useful ethical guidance requires them to understand the individual complexities of a given policy area: nobody knows enough to devise principles that are both of general applicability and also specific enough to be able to wisely guide practice.

This book exemplifies an approach to incorporating values into policymaking in the sphere of health, and in so doing aims not only to reflect on the realm of health, but to open up possibilities for others to do similar things in different policy areas. Health policy is a good place to start from in understanding the role of values in policy, as it is an area in which ethical values are difficult to overlook. As the next sections explore, there is already a large and significant literature in medical ethics and public health ethics, and a good number of national and international bodies whose aim is to set and to monitor ethical standards in the area. So someone interested in the role of values in policy has a much firmer place to start from in health, than in areas such as transport or planning policy—though I believe the approach taken here could also be helpfully extended to these and to other spheres.

1.2 Defining Health

This book above all is about what states can and should do to protect the health of their populations. Health is a rich and multivalent concept, and there are significant disagreements about the boundaries of health, and how it relates to other related concepts such as well-being, illness, disease, and disability (Boyd 2000).

Many doctors and health officials remain committed to a narrow biomedical conception of health. On the biomedical model, health is considered to be the absence of disease, and medicine a science that aims to prevent disease and repair bodily malfunctioning. Such a view is often associated with the idea that disease can be defined in a value-free way: that what disease is, and whether a given individual is healthy or diseased can be determined in purely scientific terms (Boorse 1977, 1997).

At the other end of the spectrum, the World Health Organization (WHO) has long advocated an all-encompassing definition of health as 'a state of complete physical, mental and social well-being and not merely the absence of disease or infirmity' (World Health Organization 1948). This broader conception of health has also been taken up in human rights documents such as the International Covenant on Economic, Social and Cultural Rights, Article 12, and is accompanied by a focus on entitlements not just to healthcare, but also to the conditions necessary for physical, mental, and social well-being. This definition is fairly obviously evaluatively committed, and cannot be defined in purely scientific or value-free terms.[5]

Any conception of health will involve drawing some things inside and others outside of the boundary. Depending on which account of health is adopted, adequately protecting and promoting the health of the population may require a narrower or a broader range of activities. Where does the truth lie about what health is? This may be too simple a question to ask. A better question is what we want to use the concept of health *for*. There are reasons to think that the kinds of social uses that we often want to put health to cannot be fulfilled unless we have an account of health in which health involves some kind of normative standard of functioning over and above mere statistical regularity, and hence that value-free accounts of health are not useful for policymakers (Kingma 2007). So the relevant question is which values should inform our conceptions of health and heath measurement and how they should do so, rather than whether health and health measurement should be evaluative at all. Given the variety of legitimate purposes for which health might be defined and measured, it may not be fruitful for public policy to commit definitively to one particular account of health, and of how to measure it. In order to see why, it is important to lay out some of the choices that need to be made in conceptualizing health.

It is helpful to distinguish three ways in which we can think about departures from health: disease, illness, and sickness. Disease is primarily a biomedical concept, and is most often thought of in terms of a biomedical dysfunction or a departure from a statistical norm. Whether a particular person has a disease is a question on which we would expect healthcare professionals to take the lead. Illness is indexed to the perspective and experiences of the patient, involving a feeling of pain, uneasiness, suffering, or otherwise not feeling at home in their body. Whether someone is ill is something on which the person themselves would usually be expected to be in a privileged place to determine.

Sickness involves the assignment of certain social agreed roles, in virtue of a person's ill health. Whether someone is sick is determined from the perspective of society. The sick person is often either excused from some usual duties, or given extra support, and may have usual rights or privileges removed from them. For example, the social role accorded to a sickness might entail that the individual should be signed off from work and receive a welfare benefit; or be prevented from driving. The social roles given to different diseases and illnesses are

---

[5] In practice, despite its official definition, the WHO frequently defaults to more easily tractable measures of health, such as life expectancy or numbers of fatalities, for the purposes of benchmarking the success of policies. I say more about measurement of health in the context of health related inequalities in Chapter 9.

culturally variable, and will also shift over time—something that Samuel Butler explored vividly in his 1872 dystopian novel *Erewhon*, imagining a country in which ill health was accorded the social role we would associate with criminality, and wild immorality with the social role we might associate with ill health.[6]

There are some cases—the cases that many people may think of as paradigm cases of ill health—where disease, illness, and sickness coincide. Take the case of a broken leg, or a bout of flu. In these cases, an organ or part of the body is not functioning biomedically as it usually would (the leg is broken; or the person has been infected with the influenza virus). The person themselves is in pain, and does not feel well, and there is also a recognized sick role accorded to these diseases (for example, a person would not be expected to turn up to work with flu; others would be expected to offer up their seat on the bus to someone with a broken leg).

Where a condition combines only two, or only one of these, three elements, then things are more ambiguous. For example, someone may feel ill, but if their symptoms are medically unexplained, then a doctor will not be able to diagnose a particular disease. If there is no wider social recognition for these symptoms then they may also be unable to receive a designated sick role through activities such as prescription of treatments, or provision of sick notes.[7]

---

[6] 'This is what I gathered. That in that country if a man falls into ill health, or catches any disorder, or fails bodily in any way before he is seventy years old, he is tried before a jury of his countrymen, and if convicted is held up to public scorn and sentenced more or less severely as the case may be. There are subdivisions of illnesses into crimes and misdemeanours as with offences amongst ourselves—a man being punished very heavily for serious illness, while failure of eyes or hearing in one over sixty-five, who has had good health hitherto, is dealt with by fine only, or imprisonment in default of payment. But if a man forges a cheque, or sets his house on fire, or robs with violence from the person, or does any other such things as are criminal in our own country, he is either taken to a hospital and most carefully tended at the public expense, or if he is in good circumstances, he lets it be known to all his friends that he is suffering from a severe fit of immorality, just as we do when we are ill, and they come and visit him with great solicitude, and inquire with interest how it all came about, what symptoms first showed themselves, and so forth,—questions which he will answer with perfect unreserve; for bad conduct, though considered no less deplorable than illness with ourselves, and as unquestionably indicating something seriously wrong with the individual who misbehaves, is nevertheless held to be the result of either pre-natal or post-natal misfortune' (Butler 1910, ch. 10).

[7] For completeness, there are six types of ambiguous case that can be isolated, though it is important to notice that individual conditions can and do shift between these categories as social attitudes change, or as biological knowledge increases (Hofmann 2016).

Disease without sickness or illness. Many will have had the experience of visiting the doctor for a check up to discover that, despite feeling fine, they have mild hypertension, or somewhat raised cholesterol, and being prescribed drugs or lifestyle changes because of this. Assuming the condition is diagnosed as mild, there will be no associated sick role to go with the diagnosis; and so the person may wonder what to do with the claim that they have a disease.

Disease and sickness without illness. A patient in the early stages of cancer may not feel unwell, and may be genuinely surprised to discover they have the disease. Once there is a diagnosis, the person will usually then be given a different social status as an early-stage cancer patient; for example receiving time off work for chemotherapy if necessary.

Illness without sickness or disease. A patient may feel ill, but if their symptoms are medically unexplained, then a doctor will not be able to diagnose a particular disease. If there is no wider social recognition for these symptoms then they may also be unable to receive a designated sick role.

None of the categories of disease, illness, or sickness is static; and the boundaries shift over time in response to social changes. The distinction between so-called 'normal' and 'pathological' grief provides a good example. Grief is a highly aversive experience, and so would count as an illness. As it is understood to be a universal and usually a temporary experience, doctors have not usually thought that the kind of withdrawal from the world that is embodied in grief is correctly classified as a disease. However, if the experience continues for significantly longer than is thought culturally appropriate, or if the same feelings of pain and withdrawal are present without a culturally appropriate precipitating cause, then the social role accorded to it tends to shift from a normal to a pathological one, as the condition is conceived to be depression rather than 'normal' grief. The change from the fourth to the fifth edition of the *Diagnostic and Statistical Manual of Mental Disorders* (DSM-4 to DSM-5) showed how these boundaries can shift: while there was previously a 'bereavement exemption' to diagnoses of major depressive disorder, this was dropped in one of the most controversial changes (Pies 2014).

The universal experience of ageing provides another complex example. It is uncontroversial that ageing leads to functional declines and ill health in various ways. But limitations and decline that are usual for human beings are often not considered to be disease. For example, the mean visual acuity of a 100-year-old is much less than that of a 20-year-old; but should this loss of visual acuity be thought of as a disease, or merely as the inevitable result of ageing? Alternatively, should ageing itself be thought of as a kind of deadly disease, for which we must seek a cure?[8]

In addition to the nuances introduced by separating out disease, illness, and sickness, the circumstances in which ill health arises, and its relationship to disadvantage and suffering needs to be understood contextually and ecologically,

---

Illness with disease, but without sickness. A patient may feel ill, and doctors may recognize that the disease is genuine. But there may be no relevant social role or support mechanisms in place, perhaps because the disease is thought to be minor, or so common that it is not thought to be out of the ordinary, or so rare that it does not fit into any socially recognized framework.

Sickness and illness without disease. A patient may experience an incapacitating condition, but doctors may not be able to find an adequate physiological explanation for it; alternatively, the condition may be recognized but not be thought to be a disease (for example, morning sickness in pregnancy). There may nonetheless in both cases be a recognized sick role for the person.

Sickness without either disease or illness. A condition may be defined as a disease or dysfunction from the perspective of external observers, but where the person or group labelled in this way do not in fact suffer from impaired physiological function, and do not take themselves to be ill. It is now widely appreciated that classifying a difference as a disease can be, among other things, a form of moralism or of social control. The medicalization of homosexuality now looks like a fairly clear example of such a case, but it is deeply controversial which, if any, contemporary disease labels fall into this category.

[8] Perhaps we will come to regard ageing in this way if and when there is some kind of intervention that is cost-effective enough to be made available for mass consumption. Transhumanists sometimes derisively label the idea that ageing and death are inevitable as 'deathism'. For one interesting transhumanist take on these issues, see Bostrom (2005).

through understanding our place in social and natural environments. For example, the human immune system and our capacity to function depend on a symbiosis with bacteria in the gut and elsewhere. As Chapter 10 explores, the prevalence of communicable diseases may depend as much if not more on their ecology as on conventional disease control measures. Pathogens themselves alter as a result of selection pressures—for example, mutating to become more or less virulent, or in a way that confers resistance to drugs to which they had previously been susceptible. Where a disease has a vector or vectors, these themselves add further systemic complexity to understanding disease trajectories. The combined effect of these factors is that the behaviour of individual human beings is only one element in a much larger web of interlocking causes that jointly determine the scale and the deadliness of a particular outbreak.

Not only is the causation of disease ecological, but how bad a particular depart-ure from full health is for the person who has depends deeply on socially variable factors. It has long been argued by proponents of the social model of disability that the disadvantage individuals with impairments experience is due mainly to the way that social environments are organized. The extent to which a particular impairment such as deafness or blindness impacts on an individual's ability to live the kind of life they have reason to value depends massively on the possibility of supportive technologies and supportive social structures. For example, deafness was not a disadvantage in Martha's Vineyard a hundred years ago where, as a result of several generations of congenital deafness, nearly everyone spoke sign language (Groce 1985). Being in a wheelchair has a much greater impact on one's ability to flourish in a place where nothing is disabled friendly to one where much is set up for wheelchair access (Allotey et al. 2003).

Overall, while the nature of health, and related questions about how to meas-ure health for the purpose of public policy, might initially look like factual or sci-entific questions, on closer inspection the objects of our inquiry are at least partially evaluative and socially constructed. Reflection on the ways in which the modes of ill health are constructed and contextual should make us sceptical of the idea that there is a single correct account of health or of how to measure it.

The questions about the role of the state in protecting and promoting health which this book focuses on are informed by these debates about the nature of health, but for two reasons are not fully determined by them. First, much of this book will focus on paradigm cases of ill health, which bring together disease, ill-ness, and sickness. There will be some cases where it is clear that a broader dia-logue about health, illness, and disease, and how to measure them, will be required, but we shall note these as we encounter them.

Second, competing conceptualizations of heath may make a significant differ-ence to whether a phenomenon such as social isolation should be conceived of as a *health* problem, but it is less obvious that shifts in the boundaries of the concept of health should make an equally significant difference to a government's duties to its citizens. Ensuring the conditions in which citizens can achieve a decent level of well-being seems to be an important goal for governments, regardless of whether health is conceived of as synonymous with well-being, or in much narrower terms. While the standard structure of government departments, with each having separate responsibilities, makes sense as a response to the complexity and multifariousness of governments' duties, it would be a mistake to think that whether social isolation is classed as a health as opposed to a social care or a transport problem should make a fundamental difference to duties to ameliorate it. Rather than treating something as a problem for government to solve only if it fits neatly into a pre-existing departmental silo, governments should start from individuals and the kinds of lives they are able to lead, and work back from there to how to structure governmental responses.

1.3   The Idea of a Public Health Problem

I use the term *public health* to refer to a kind of activity that is collectively undertaken to ensure that the health of the population is protected or promoted (Verweij and Dawson 2009). Public health is concerned classically with interventions that are collective in two senses: they involve the concerted action of a number of people or institutions in order to bring them about; and second, they aim to improve health at a group or population level, as opposed to merely an individual level. Public health should be contrasted with *population health*, which I understand as a measure of the level of health of a particular group.

Given the doubly collective nature of public health interventions, there are many ways in which individuals can intervene to improve their own health, or the health of others, that do not amount to public health. Ordinary clinical medicine, in which one (or a number) of healthcare professionals are aiming to benefit a single patient, does not count as public health; nor does a particular individual's decision to stop smoking. The broader decision whether to have a rotavirus vaccination campaign, or a campaign to promote smoking cessation, is a public health decision.

The boundaries between public health and individual-level health interventions are porous and in flux, particularly in a system such as that in the UK or Canada where the vast majority of individual-level health interventions occur within a publicly funded healthcare system that is shaped by and integrated with public health goals. As will become apparent as the book progresses, the net effect of my view is to rethink ordinary clinical care through the lens of public health. Rather than thinking of public health as something different from and needing to be kept conceptually separate from clinical care, on my view public health is the normative foundation within which clinical care operates.

To describe a condition or a behaviour as a public health problem is to imply two things about it. First, that it significantly reduces population health (however population health is understood), and second, that this reduction in population health should be counteracted by public health measures. The latter point is crucial, as it is quite possible to agree, for example, that obesity leads to a significant reduction in population health while being sceptical that reducing this population health deficit should be a target for public health activity (Anomaly 2012).

There are two kinds of reasons given for excluding things that are admitted to cause reductions in population health from being classified as public health problems. First, risks to health can be acknowledged as such, and seen as a price worth paying by those who are taking them. If those who assume a risk to their health *prefer*, say, to drink a bottle of wine a day and have a reduced life expectancy, to a longer and soberer life, then it might be argued that it is no business of the state to force them to be healthy, and that it is illegitimate to describe the collective effect of the free choices of individuals, which tend to reduce population health, as a public health problem.

We examine this type of argument at length in Part II, and reject it both because of doubts about whether such health damaging behaviours are chosen in a way that removes any duty for the state to act, and also more broadly because there can be duties on the state to reduce health risks even where those risks are voluntary under some description.

Second, it is sometimes argued that the role of the state should be limited, and that the nature and legitimacy of public health should follow from these limits. This is usually linked to a particular conception of the role of the legitimacy of state action: namely that it is difficult to justify state interference in individuals' liberty unless there are specific features such as the fact that the intervention is needed to ensure supply of a public good, or to prevent violation of rights. One prominent way in which this argument has been advanced is to claim that public health aims only at (or ought to aim only at) the provision of public goods—benefits that are nonrival and nonexcludable (Anomaly 2011, 2015; Horne 2019).

This seems to me to be mistaken both empirically and normatively. It is true that some goals of public health policy (such as the eradication of polio) are paradigmatic cases of public goods. But eradication is at the very far end of a continuum, and most vaccination campaigns have a significantly different profile. Some vaccination campaigns (such as measles vaccination) aim to establish the public good of herd immunity as well as protecting individuals. Others (such as seasonal influenza in England) have as their main aim giving protection to the most vulnerable, and do not aim to establish sufficient coverage to ensure herd immunity. Lastly, there are vaccination campaigns for tetanus, which is not communicable from person to person and so the vaccination cannot create a public good. It strikes me as implausible to think that while measles vaccination is a public health activity, seasonal influenza or tetanus vaccination is not.

I am thus inclined to think that the idea of public goods is a red herring when it comes to defining public health. The core of public health lies in systematic attempts to reduce health risks, rather than in the provision of public goods. There is no requirement to show that a problem has the structure of a public good problem or another classic case of 'market failure' before it is legitimate for the state to intervene: what matters is the ethical importance of the goods that can be secured by so doing (Dees 2017). I thus adopt an inclusive conception of a public health problem: any health problem could be thought of as a public health problem, so long as it has a population-level impact, and is amenable to action at a large scale.

1.4 The Context of Public Health Ethics

Medical ethics is a sub-branch of ethics, which applies ethical analysis to medicine and healthcare. Medical ethics in the modern sense (and the broader discipline of bioethics) were both were born in the USA at the beginning of the 1960s. The types of ethical problems that early bioethics focused on, and the approaches it took to them, were shaped by certain features of wider societal change in the 1960s and 1970s. The net effect was that despite the fact that public health measures by their nature tend to raise ethical questions, ethical discussion of public health was largely neglected within this literature until the turn of the millennium. There is a longer story to be told, but for present purposes, two features are particularly important.

First, the birth and growth of bioethics coincided with a period of great optimism about the future of medicine. The wide availability of penicillin from 1945 onwards, and the subsequent rapid progress in discovering new antibiotics, revolutionized medicine (Le Fanu 2000). At the same time, implementation of effective vaccination regimes brought many major killers such as tuberculosis and polio under control, while smallpox was eliminated entirely. Innovations such as heart transplants and IVF became available. The assumption was made that such new technologies were the key to the future of medicine, and that communicable diseases no longer posed a long-term threat.[9]

---

[9] A widely quoted, but possibly apocryphal, statement by the US Surgeon General, William Stewart has come to emblematize this mood of optimism: 'The time has come to close the book on infectious diseases. We have basically wiped out infection in the United States.' This statement is said to have been made around 1970, though as Battin et al. (2008) relate, it is difficult to find an original source for it. As a result of this attitude, there was very little early bioethics literature on communicable disease, with the exception of HIV/AIDS which tended to be treated as a special case.

Second, societies became more individualistic, and there was a much greater willingness to challenge traditional forms of authority. As society became more individualistic, the traditional doctor–patient relationship began to be questioned. Patients became much more vocal, and wanted more control in their dealings with doctors. A key question became how to defend the autonomy of patients against the power and expertise of doctors (Katz 2002).

Given this background, it is unsurprising that key themes in the early years (1960s–2000) of bioethics were informed consent, the ethics of research (spurred on by scandals such as the Tuskegee syphilis experiment), and the 'Promethean challenges' created by new technologies such as organ transplantation and IVF (Daniels 2006). Work tended to focus on the doctor–patient relationship, and this was thought of in individualistic terms, without much attention to the broader social structures within which medicine was being practised. Particular features of US society—such as its high degree of individualism—plausibly increased this focus on individuals at the expense of wider social context (Holm 1995).

Since the turn of the millennium, this type of bioethics and the assumptions that lay behind it, have been comprehensively undermined both by social change, and by new forms of interdisciplinary scholarship that have shown that health is inextricably linked to broader decisions about the structure of society. We can group these under four headings: ageing societies and the increasing prominence of chronic disease, the social determinants of health, the rising cost of healthcare, and the return of communicable diseases.

1.4.1 Ageing Societies and the Increasing Prominence of Chronic Disease

Past decades have seen a sharp rise in life expectancy. This rise in life expectancy has both been caused by shifts in patterns of mortality and morbidity worldwide, and as populations have aged, it has itself led to an increase in long-term conditions such as dementia and type 2 diabetes. From 1970 to 2016, global male life expectancy at birth increased from 56.4 years to 69.8 years and global female life expectancy at birth increased from 61.2 years to 75.3 years. Global life expectancy at birth rose by 3–4 years every decade from 1970, apart from during the 1990s (when there was an increase in male life expectancy of 1.4 years and in female life expectancy of 1.6 years) (Wang et al. 2017).

One of the most significant drivers of this shift has been a fall in maternal and under five mortality, and in deaths from communicable diseases; and a consequent rise in the number and percentage of deaths from non-communicable diseases. In 2016 communicable, maternal, neonatal, and nutritional causes were 19.3 per cent of total deaths worldwide, which was down from 34.1 per cent

(15.9 million of 46.5 million) in 1990. Meanwhile, deaths from non-communicable diseases rose by just under 8 million between 1990 and 2016, accounting for 72.3 per cent (39.5 million) worldwide by 2016 (Wang et al. 2017).

As societies have aged, many more are living with chronic diseases that need to be managed—whether through individual care or broader public health measures. Chronic diseases are by definition at present not curable. Where cure is not available, medical success is better understood in terms of maintaining or expanding the patient's well-being or capability to do the things they care about, than purely in terms of maintaining biomedical functioning. Chronic disease patients are usually also their own main caregivers. They will usually see healthcare professionals for periodic check-ups, but this may be no more than a few times a year. In between those times, they will attempt to live their lives while following their care plans. What makes the biggest difference to how well the condition is managed is what the patient does and what happens to them in the vast majority of the time that they are outside of a clinical environment. This has required a broadening of the focus of bioethical reflection away from the doctor–patient relationship (Walker 2019).

1.4.2   The Importance of the Social Determinants of Health

Much more is required to ensure that everyone has the conditions in which they can live a long and healthy life than simply providing access to healthcare for all on the basis of need. Indeed, equal entitlement to comprehensive healthcare is compatible with massive differences in life expectancy—as, for example, was highlighted by the 2008 WHO Commission on the Social Determinants of Health, which revealed a 28-year difference in male life expectancy between the richest and the poorest parts of the Scottish city of Glasgow (Commission on the Social Determinants of Health 2008).

Healthcare is thus only one of a number of determinants of health. Other determinants include workplace stress, social exclusion, early nurturing environment, and the inequality of distribution of income in society, and are highly correlated with socioeconomic status (Wilkinson and Marmot 2003). Moreover, the ability to take advantage of the full range of healthcare services provided, and to change behaviour in response to new health information, is itself highly correlated with socioeconomic status. The uncomfortable implication is that even policies which successfully improve population health will often end up making inequalities in health worse. Smoking provides a powerful example: before Doll and Hill's groundbreaking 1954 study (Doll and Hill 2004) on the relationship between smoking and cancer, smoking was extremely widespread in all social classes. When the health risks of smoking became apparent, smoking rates declined much more steeply amongst the rich than among the poor, widening the gap between the health of the rich and the poor.[10]

1.4.3 Rising Costs of Healthcare

The trajectory of healthcare spending in developed countries has come to be seen widely as unsustainable. For example, when the UK National Health Service was founded in 1948, it had a budget of £437 million. This would be roughly £15 billion at today's value. For 2016–17 it was £144 billion. In other words, adjusting for inflation, the UK was spending ten times as much on health services as it did 70 years previously at the founding of the NHS. Public spending on health in the UK has been growing at an average of 3.7 per cent per year in real terms since 1948 (Harker 2018).

Part of this increase has been offset by economic growth in GDP—but the fact remains that in 1949–50 the UK government spent roughly 3.5 per cent of GDP on health, and in 2011 it spent around 7.3 per cent (Stoye 2018). This figure rises to 9.6 per cent of GDP including non-government spending (Office for National Statistics 2019). Many other countries spend a great deal more than the UK: Germany spent 11 per cent of its GDP on healthcare in 2011, and the USA spent 17 per cent (World Bank 2019a).

This vast increase in healthcare budgets over time has complex and mutually interlocking causes: in part it is due to the epidemiological shift implied by ageing societies and the increased prominence of long-term conditions such as dementia, but it is also due to the benefits provided by medical research, and the often high price of new pharmaceuticals. For example, in 1939, the median age of survival of children with cystic fibrosis (CF) was six months (Davis 2006), but by 2015 it was over 40 years (Keogh et al. 2018). By 2015, lumacaftor-ivacaftor (Orkambi) had been launched, promising significant improvements in length and quality of life for around half CF patients—but at an extremely high cost of $272,000 per patient per year.

Similar stories of great achievements that significantly increase costs occur repeatedly in healthcare. In short, while the level and quality of healthcare that could be provided is improving all the time, and the amount that could be spent is also increasing, costs are rising faster than budgets. This leaves a set of very difficult problems about how best to allocate healthcare resources in an environment where we will probably not be able to meet all healthcare needs.

Rising costs, and in particular costs due to long-term conditions that cannot be cured but only managed, invite increased attention to primary, rather secondary

---

[10] Within the UK, by 2016, rates of smoking had fallen to 7.9 per cent in the least deprived decile, but remained at 27.2 per cent in the most deprived decile. See Office for National Statistics (2016).

and tertiary prevention. Primary prevention aims to reduce the likelihood of disease or injury before it occurs (for example, antismoking campaigns). Secondary prevention aims to diagnose disease early in order to allow interventions that will minimize the effects (for example, cancer screening; training of employees to spot the signs of workplace stress). Tertiary prevention aims to minimize the effects of disease or injury that is already severe enough to have made a noticeable difference to the patient's life (for example, rehabilitation programmes after a stroke; antiretroviral drugs after an HIV diagnosis). This book argues both that primary, secondary, and tertiary prevention should be thought of as part of the same integrated system, and also that potential health gains from secondary and tertiary prevention need to be weighed in prioritization decisions against the (often more cost-effective) health benefits that can be gained from primary prevention.[11]

1.4.4 The Return of Communicable Diseases

It is vital to give due emphasis to the rise of long-term conditions, but it is important also to balance this with a recognition that the hope that communicable diseases had been vanquished was always a pipe dream. The resurgence of tuberculosis, the rapidly increasing threat from drug-resistant infections more generally, and the rise of emerging diseases such as Ebola showed that bioethics had unjustly neglected communicable disease, even well before the Covid-19 pandemic (Selgelid 2005; Littmann and Viens 2015). Multidrug-resistant tuberculosis alone now accounts for over 150,000 deaths per year (World Health Organization 2018b).[12] The 2016 O'Neill Review on Antimicrobial Resistance estimated that drug resistance infections already caused 700,000 deaths per year globally, and estimated that this could rise to 10 million per year without action (Review on Antimicrobial Resistance 2016), a threat that then WHO Director General Margaret Chan described as a 'slow-motion tsunami' (Chan 2016).

Looking at infectious disease requires a broadening of perspective from that of the individual doctor–patient relationship, for two main reasons. First, the conditions that are conducive to the spread of infectious disease are closely related to poverty. The main reason for the continued size of the burden of communicable disease is a collective failure to secure the basic prerequisites of healthy life for more than a billion of the world's poorest people: clean water, access to improved sanitation, adequate nutrition, shelter, air quality, and health facilities.

---

[11] Chapter 7 discusses these questions at much greater length.

[12] Even in the EU, multidrug-resistant bacteria accounted for an estimated 33,100 deaths in 2015 (Cassini et al. 2019). This is likely an underestimate due to a frequent lack of recording of antimicrobial resistance on death certificates.

Second, there is an important difference between a noncommunicable disease such as diabetes, and communicable diseases such as tuberculosis or influenza. The person with the communicable disease, in addition to being a patient, also presents a threat to the health of others, and such threats if left unchecked may lead to untold numbers of new cases. In the name of one book, the patient is both a victim and a vector of the disease (Battin et al. 2008).

1.4.5 Systemic Interconnections and Clustering of Risk Factors

The first two decades post-millennium thus saw a pervasive shift in the focus of medical ethics and bioethics away from an almost exclusive focus on what goes on between doctor and patient, or between researcher and research participant, towards a more socially engaged and complex discourse that treats potentially all the factors that affect health in a population as fair game for ethical analysis.

These four elements—ageing societies, the social determinants of health, rising costs of healthcare, and the return of communicable disease—should not be seen as separate, but as interlocking challenges, given that risk factors for disease tend to cluster and to follow a social gradient. To return to the example with which I started the chapter, obesity is a predisposing and exacerbating factor for many other diseases. Obesity rates are highly correlated with social status in most countries: the lower the social status, the higher the rates of obesity. Obesity can have further synergistic interactions with other social factors such stigmatization, which both worsen health and reduce healthcare-seeking behaviour. In this, and in many other cases, clustering of risk factors can reinforce one another, reduce well-being and make successful intervention more difficult (Powers and Faden 2008: 71–8).

Appreciating the importance of systemic interconnections requires a shift of emphasis for ethical reflection and for public policy. As Chapter 2 examines in greater depth, approaches that may have looked like the future of medicine and of public health, in which specific interventions were based on randomized controlled trials, may need to be revised and replaced with approaches that prioritize complex systems.

1.5 A Brief Map of What Is to Come

This book advances an account of how and why philosophy should inform the foundations of public health policy. Doing so requires policymakers to rethink the nature of public health policy, but also philosophers to rethink the nature of philosophy. Part I, 'Philosophy for Public Policy' argues for a re-orientation of philosophy towards solving real-world problems—a philosophy which acknowledges that the ethical problems we face are historical and contextual, and that solving one problem may often contribute to worsening others.

Chapter 2 examines the role of evidence in public policy. Simplified testing environments such as randomized controlled trials (RCTs) are often thought to provide the most rigorous way of establishing causal claims. On this basis it has been claimed that what public policy requires is a solid evidence base of RCTs, which are then synthesized into an account of 'what works'. However, external validity does not follow from internal validity. Even if a trial is internally valid—that is to say, that it can be shown with confidence that the intervention had a particular size of effect within the trial population at a particular time—this does not imply external validity, namely that the intervention will work in a wide range of contexts, and in particular in our context. In order to be confident that interventions can be transferred successfully from one context to another, researchers usually need to understand a good deal about the support factors that allowed the intervention to be successful where it was, the extent to which their own context is different from the experimental environment, and how to modify the intervention or support factors in the light of these differences. A number of factors, including the greater difficulty of controlling for confounding factors, and the greater variability in support factors, make external validity much more difficult to achieve in public policy than in clinical medicine.

In Chapter 3, I argue that the distinction between internal and external validity applies also to ethical reasoning. A particular approach to ethical reasoning has recently come to dominate much Anglo-American philosophy—an approach that assumes that the most rigorous method is to proceed by analysis of thought experiments. In these thought experiments, features such as context and history are stripped away, and all factors other than those of ethical interest are stipulated to be equal. This approach faces exactly the same problems that RCTs do. Even if a thought experiment produces results that are internally valid—in that it provides a genuine ethical insight about the highly controlled and simplified experimental scenario under discussion—this does not imply external validity. Just as in empirical experiments, there is a yawning gap between succeeding in the relatively easy project of establishing internal validity in a controlled and simplified context, and the more difficult one of establishing external validity in the messier and more complex real world. Yet the problem of external validity in ethics is one that few philosophers have even noticed, let alone begun to solve.

Chapter 4 argues that the scale of the challenge posed by external validity requires a similarly sizeable response. Not only should practitioners approach evidence collection and interventions in policy differently, but also philosophers should change the way they conceive of ethics. The default should no longer be to start from simplistic causal models or thought experiments, while being dimly aware that these approaches will exclude some features that would be relevant for real-world decision-making. Rather, both practitioners and philosophers should start from the premise that social processes are complex systems. On a complex systems approach, creating and testing models is essential for responsible interventions, but there will also be a deep acknowledgement that models are by their nature simpler than the systems they are trying to model, and thus model predictions can and will diverge in important ways from what actually happens. Moreover, complex systems are in important aspects performative: for example, what counts as a breach of trust or a violation of privacy is not something that can be discovered once and for all, but is partly constituted by social norms and individual expectations, which will themselves change in response to government action. Taking performativity seriously means rethinking the assumption that there is a static ethical reality, which ethical concepts and theories should be attempting to map accurately.

Part II, 'Beyond the Neglectful State: An Ethical Framework for Public Health', examines the role of ethical values in public health. The analyses of Part II and Part III of this book presuppose a particular context: middle- or high-income democratic states, which have an effective government, and a commitment to treating their citizens as equals. These are states that are receptive to—even if not yet fully signed up to—the claim that protection and promotion of population health is not only a legitimate goal for the state, but a potentially important one. These are also states that take the liberty and autonomy of individuals sufficiently seriously that they acknowledge a duty to justify government action where it would involve significant interference with autonomy or reduction of liberty. I take it that this description applies to all, or nearly all, of the states in the OECD, and to a significant number outside it.

Chapter 5 analyses the extent to which it is problematic for the state to improve health by interfering in individuals' lives. It argues that complaints about paternalism in public health policy are much less convincing than is often thought. First, it is deeply problematic to pick out which policies should count as paternalistic; at best we can talk about paternalistic justifications for policies. Second, two of the elements that make paternalism problematic at an individual level—interference with liberty and lack of individual consent—are endemic to public policy contexts in general and so cannot be used to support the claim that paternalism in particular is wrong. Instead of debating whether a given policy is paternalistic, we should ask whether the infringements of liberty it contains are justifiable, without placing any weight on whether or not those infringements of liberty are paternalistic. Once we do so, it becomes apparent that a wide range of interventionist public health policies are justifiable.

Chapter 6 argues that public health needs to be rethought through a rights framework. There is a right to health, and this entails that individuals have a right to public health. I argue that while interventionist public health policies are often accused of being paternalistic, or to show the 'Nanny State' in action, things are actually rather more complex. Given that there is a right to public health, we also need to guard against something arguably even worse: the Neglectful State. The Neglectful State is one that does not take easy steps that could have been taken to reduce risks to health, and as a result allows significant numbers to come to harm or die. The ethical challenge of public health policy is therefore not the one-sided one of avoiding Nannying, but the more complex task of steering a course between Nannying and Neglect. Avoiding Neglect may involve restricting liberty in various ways. Though it might seem paradoxical to argue that an individual right to public health could justify restricting liberty, the chapter argues that this is an accepted implication of other rights such as the right to security, and so that there is in fact no paradox.

Chapter 7 addresses how to prioritize public health policies within the context of the right to health approach. The touchstone of this approach is that public health interventions need to be justifiable to individuals. It is not enough, for example, to merely cite the size of the population health gain, if the intervention is one that involves an interference that would seem disproportional to most of those it affects. Designing approaches to prioritization that are adequately justifiable to individuals can be extremely difficult. One tool for clarifying the problem, which has been explored widely in the philosophical literature, is the idea of a *claim*—where the strength of an individual's claim depends on features such as how badly off they are, their capacity to benefit, the time at which their need arises, and whether the bad that will befall them is certain or merely possible. The chapter argues that it is mistaken to think that there is a single and uniquely correct way of measuring claims. Thus, approaches to prioritization need to be pluralistic, and need to reflect on the measures most appropriate for a particular policy challenge.

Part III, 'Structural Justice', integrates the complex systems analysis developed in Part I with the ethical framework developed in Part II. It examines three spheres in which public health policy needs to make choices—responsibility, equality, and networks, focusing on three challenges: (1) how to make use of judgements of responsibility, and whom to hold responsible; (2) how to specify the goal of health equity and how to pursue it; and (3) the implications of the fact that most health risks are contagious or can be amplified by socially mediated networks of causes.

Chapter 8 examines critically the idea of responsibility, arguing that which aspects of an individual's condition they can reasonably be held responsible for should be determined by the values that the health system is aiming to promote or respect, rather than by treating personal responsibility as an extrinsic ethical requirement on health system design. A health system's answer to the question of whom to hold accountable, and how to do so, thus needs first of all to be framed within the context of the right to public health. In so far as claims of irresponsibility can be fairly levelled, these should in the first instance be directed towards those who violate the right to public health, either through government or corporate agency, rather than at isolated individuals. Health systems should aim to expand and protect individuals' effective ability to *take responsibility* for their health, but it will usually be the case that *holding these individuals accountable* (or threatening to do so) will not be a very effective means of so doing. Acknowledging the importance of individuals taking responsibility for their own health is consistent with a resolute resistance to blaming or otherwise holding to account those who for one reason or another do not do so effectively.

Chapter 9 examines how health systems should measure, and respond to, health-related inequalities. Health equity is often taken to be a core goal of public health, but what exactly health equity requires is more difficult to specify, for two reasons. First, there are indefinitely many health-related variables that can be measured, and variation in each of these variables can be measured in a number of different ways. Second, given the systemic interconnections between variables, making a situation more equal in some respects will tend to make it less equal in others. The chapter argues that a number of existing philosophical approaches are too simple: they tend to assume that there will be only one type of health inequality that really matters, and also tend to assume that it is inequality per se that is the problem rather than features which accompany inequality. The chapter argues that it is better to adopt a pluralist approach to health equity measurement. What measures of health equity need to be accountable to is the lived experience of individuals' lives and the ways in which power impacts on these. Reflection on the deepest and most resilient causes of health-related inequalities shows that they are often the result of intersecting structural concentrations of power—structures that it is vital, but very difficult, to break up.

Chapter 10 examines the idea of contagion—of risk magnification and modulation through networks. While this is most familiar within public health in the context of communicable disease, adopting a complex systems approach makes salient that networks matter for public health in other ways too: social norms, food cultures, and attitudes towards body shape are all shared and amplified through social networks and are thus contagious. The chapter examines three case studies, which each raise different questions about the interplay of causal complexity, performativity, and policymaking: vaccination policy, drug resistant infections, and disease eradication. In vaccination policy, achieving herd immunity is often crucial, but attempts to do this are heavily dependent on public trust. Drug resistant infections arise, among other causes, through the inevitable impact of natural selection, and so require a shift towards an ecological perspective on disease. Finally, the possibility of disease eradication poses important questions about when and how to ensure that susceptible health threats are systematically and permanently removed from the environment.

PART I

PHILOSOPHY FOR PUBLIC POLICY

2

Evidence, Mechanisms, and Complexity

2.1 Introduction

Public policy often fails. It is common, rather than rare, for policies that are designed to solve a problem to either have little impact on improving it, or to have unanticipated effects that worsen people's lives. Part I of this book aims to explain why this is, and to offer a way forward.

Two basic questions can be asked of any problem that is significant enough to require government attention. First, what are the mechanisms that are responsible for creating or sustaining this problem? Second, what intervention or interventions should be deployed to remove or ameliorate it? There are a number of ways in which a policy can fail to achieve the goals it is designed to achieve. Most obviously, policymakers may be wrong about the relationship between an intervention and the behaviour or problem they are aiming to influence. For example, they may assume that abstinence-only sex education will be an effective way of reducing rates of teenage pregnancy, but may be wrong about this (Stanger-Hall and Hall 2011).

Policies also may be successful in their own terms, but fail to shift the dial on an underlying systemic problem. Chapter 1 explained how for public policy problems like the rise of obesity, or large and persistent inequalities in health, the mechanisms that create and sustain the problem are many, and some of these will be much more easily altered by government action than others. Even if one or more mechanisms that are responsible for exacerbating a problem are correctly identified, and a successful intervention to remove or counteract these mechanisms is performed, the underlying problem may continue to get worse, if the mechanisms targeted for intervention have an insufficiently large effect relative to the overall causal structure.

This chapter focuses on two questions about the role of evidence in public policy. What counts as good evidence that a policy will work? How transferable is evidence of success of a policy in one context to other contexts? In doing so, it tells a story about the evidence-based medicine movement, and the rise and then contestation of the idea that approaches to collection, grading, and collaboration of evidence that were popularized by evidence-based medicine will also work in policy.

As we shall see, evidence-based medicine is founded on the assumption that evidence gathered about the effectiveness of a particular intervention in one context will also provide evidence for the effectiveness of that intervention in other contexts; and that evidence gathered from the same intervention being tried in multiple different contexts can be combined to form an evidence base, which can guide decision-making much more effectively than relying on the results of any one individual intervention. While these assumptions are often accurate in clinical medicine, they are deeply problematic in the context of public policy.

In the light of these concerns, debates around the role of evidence within public policy have emphasized increasingly the importance of context and of complex systems approaches—which Chapters 3 and 4 explore. At the heart of the chapter is an insight that we will return to often in this book: progress is neither simple, nor one-dimensional. Often even genuine progress will generate new problems further down the line.

2.2 The Rise of Evidence-Based Medicine

Evidence-based medicine (EBM) began as a movement in medicine spearheaded by a research group at McMaster University, led by David Sackett and Gordon Guyatt in the 1970s. The revolution has now long since been complete. EBM is now a worldwide movement, and its fundamental ideas have been embedded into medical training, the ways in which doctors practice, and the methods used by journals to assess the quality of evidence. As the movement grew, so it broadened—so that it is now not just a movement based on reforming what doctors, or even healthcare professionals more broadly, do, but what practitioners in a wide variety of fields do. Hence, it is often now referred to as the evidence-based practice movement.

Evidence-based medicine defines itself as the 'conscientious, explicit and judicious use of current best evidence in making decisions about the care of individual patients' (Sackett et al. 1996). On the face of things, this sounds like it should be unobjectionable, and indeed obvious. It is hard to believe that those whom the EBM movement critiqued would have accepted that their practice was not evidence-based. So there is an element of persuasive definition in the way that EBM talks about 'best evidence'. Greenhalgh gives a more enlightening definition: 'Evidence based medicine is the use of mathematical estimates of the risk of benefit and harm, derived from high-quality research on population samples, to inform clinical decision making in the diagnosis, investigation or management of individual patients' (Greenhalgh 2006: 1).

Viewed in this light, there was something genuinely novel about placing at the heart of medical practice mathematical estimates of risk of benefit and harm that had been formed from high-quality research on population samples. For the vast majority of its history, Western medicine was deeply shaped by assumptions about the nature of disease that were deeply hostile to this idea. Humoural medicine, which was the dominant model from antiquity until the nineteenth century, is based on the idea that there are four humours (black bile, yellow bile, phlegm, and blood). Disease was understood to be caused by an imbalance among the humours. Each person was thought to have their own unique balance of the humours—and so medicine by its nature had to be practised in a way that required attending (and listening to) the particularities of each individual's case. Given these assumptions, doctors did not take themselves to be in need of universal principles that could be applied to all patients.

This mode of clinical medicine was unfortunately not very effective in curing disease. Indeed, it is argued that for much of the long history of medicine (up till the 1860s), scarcely any of the treatments offered by doctors was more effective than placebo (Wootton 2007). Favoured treatments such as bloodletting, emetics, and purges were positively harmful. Moreover, there was no systematic attempt to test whether the treatments offered by doctors were effective, or whether they did more harm than good.

The second half of the nineteenth century, and early twentieth, saw decisive shifts in the underlying understanding of the ontology of disease and the role of science in medicine.[1] Testing of treatments, evidence collation, and change of medical practice progressed more slowly. The first rigorous randomized control trial (RCT) was not published until 1948, when the British Medical Research Council examined the treatment of pulmonary tuberculosis with streptomycin (Medical Research Council 1948). Following this, the use of randomization for testing of treatments became standard—with the US Food and Drug Administration requiring proof of efficacy via RCTs for new drugs from 1962 onwards (Jones and Podolsky 2015).

Even so, the take-up and use of research evidence in medical practice remained patchy. There were widespread differences in standards of practice for patients with essentially the same condition (Eddy 2005: 10). Most of the procedures in use had never been rigorously tested, and even where good quality research had been published, it would take a very long time for medical practice to change. Archie Cochrane (1972) played a major role in publicizing these problems, and argued that these failures to use evidence effectively was costing lives. For example, the failure to do a systematic review of the evidence on corticosteroid administration before preterm delivery led doctors not to use the practice—when had the available data properly been collated it would have been apparent that this significantly improved outcomes (Roberts et al. 2017). The forest plot diagram which was used for this paper then became the logo for the Cochrane Collaboration library of systematic reviews, named after Archie Cochrane.

---

[1] Milestones included the germ theory of disease, aseptic surgery, Koch's postulates of infectious disease causation, and the Flexner Report's advocacy of science-based medical education.

EBM as a movement has consistently argued that evidence should be graded, and that this grading should occur on the basis of features of the design of published studies, rather than on the specifics of the physiological mechanisms appealed to or arguments developed within them. Two crucial aspects of the approach to grading have been first, to give much greater weight to randomized trials than ones that are not randomized and second, to assume that systematic reviews or meta-analyses that combine evidence from a number of different studies will be more reliable than those of single studies. EBM has also tended to downplay the importance of mechanistic and theoretical studies, which attempt to provide evidence or arguments that are relevant to which interventions will work on the basis of physiological or broader explanatory principles.[2]

As Section 2.4 explores, this downplaying of mechanisms has significant costs. Most importantly, without an in-depth of understanding of the mechanisms and causal pathways that are being targeted by an intervention, it may not be clear why an intervention works when it works, and in particular how reliable it will be in circumstances that are somewhat different from the background conditions in the intervention. Thus, there are reasons to doubt that EBM's championing of randomization represents clear progress over models of thinking about causation that were current in medicine before EBM, such as the Hill (1965) criteria for determining whether an association is causal or accidental. Hill's criteria include elements that could be tested by randomized trials, but also core elements that could only be proved or interrogated via mechanistic or theoretical studies, such as whether the proposed cause occurs prior to the effect, the causal claim's theoretical plausibility, and its coherence with existing scientific knowledge. Increasingly, thoughtful theoretical work on EBM accepts, following Russo and Williamson (2007) that hunting for causes in medicine requires both probabilistic evidence from experimental trials and mechanistic evidence if it is to be rigorous. Parkkinen et al. (2018) develop an account of how to integrate evidence of mechanisms into EBM.

2.3 From Evidence-Based Medicine to Evidence-Based Policy?

The idea that evidence-based approaches can and should be adopted for the policy realm initially seemed like a logical extension of EBM. Sir Iain Chalmers (2003), who was instrumental in the development of the EBM movement, provides a good example of an early and enthusiastic attempt to do this. He adduces a

---

[2] In a leading EBM textbook, Sackett et al. (2000: 108) once made the now infamous claim that only randomized studies are worth reading: 'If the study wasn't randomized, we'd suggest that you stop reading it and go onto the next article.' The more recent consensus GRADE guidelines still treat randomized studies as much to be preferred. See Guyatt et al. (2008); Balshem et al. (2011).

number of examples of policies that have failed: driver education programmes in schools that may increase overall death rates on the roads, as they get people driving earlier, but do not affect crash rates; 'scared straight' programmes, which aimed to prevent young men from crime by showing them what it is like to be in prison, but in fact had the opposite effect; and Dr Spock's advice that parents should put their children to sleep lying on their stomachs, which greatly increased the chance of sudden infant death syndrome (SIDS) (Chalmers 2003).

Chalmers takes the argument for evidence-based policy to be simple: given that policymakers intervene in the lives of citizens by altering the conditions in which people live, and that these policies may end up doing more harm than good, it is important for them to base these interventions on the best evidence. On this view, just as in clinical medicine, policy researchers must undertake randomized trials, because randomized trials are better at controlling for bias than non-randomized trials, and inadequately controlled studies often overestimate the size of intervention effects. As he puts it, 'Those who reject randomization are implying they are sufficiently knowledgeable about the complexities of influences in the social world that they know how to take account of all potentially confounding factors of prognostic importance, including those they have not measured, when comparing groups to estimate intervention effects' (Chalmers 2003: 30). Information on what works should be collated via systematic reviews, which will assemble the relevant evidence into a form in which decision-makers can gain an overview of it. Where there are gaps in the evidence base relevant for policy, these should be plugged by undertaking further policy-relevant randomized research.[3]

Would that things were so simple. There are a number of deep difficulties with this strategy. First, even if it were true that RCTs provide a better way of testing the effectiveness of interventions than any other method, it would remain the case that some kinds of interventions are much more amenable to test via RCTs than others, and that public policy is particularly likely to involve interventions that are difficult to test via RCT. As Chapter 9 explores, the distribution of health in a society has structural social determinants such as wealth inequality, racism, and mental health stigma, as well as more local determinants. Structural determinants of health are likely to be resistant to being altered by small-scale interventions—just in virtue of the fact that they are themselves created and held in place by powerful social forces (Sommer and Parker 2013).[4] Policy interventions

---

[3] This style of approach been advocated by, amongst others, the Cabinet Office Behavioural Insights Team; see Haynes et al. (2012). Randomized trials have also become much more popular in Development Studies; see, for example, Banerjee and Duflo (2011).

[4] This is not to say that small-scale interventions will have no effect, or that many small-scale interventions working together could not make a large difference to a structural determinant of health. The point is rather that policymakers need to anticipate the structural features that may make simple interventions ineffective, and to redesign accordingly. The journey of the Bill & Melinda Gates Foundation, from advocating the use of discrete and isolated interventions in the early 2000s, to a

that aim to make structural changes are particularly difficult to test in an RCT. There are good reasons to think that there are many interventions that would be very effective (or conversely, very destructive) but which are for a variety of reasons not testable in an RCT.[5] It is important not to confuse lack of RCT evidence with lack of a reason to think that a policy would be worth adopting.

Second, there is usually much less ability to control for confounding factors in public policy than in clinical trials, and so the data will be much messier. In a clinical trial there will be explicit control over the characteristics of who is allowed to enter the trial; the drug that will be tested is also precisely specified and will not alter between the participants; the trial will be run according to a precise protocol; and the trial will often be double blinded. Such controls are more difficult, and sometimes impossible in a public policy context. For example, if an intervention is designed to be delivered by all teachers in a district with the aim of reducing rates of felt body shame in obese pupils in the school age population, then it is far from easy to adequately isolate the effects of the intervention from other systemic effects such as shifts in advertising, broader public culture, and the effects of other interacting policies. The intervention itself is more difficult to standardize than a pharmaceutical intervention, because the take-up of information-giving and advice is partly dependent on features such as the level of trust in the person giving the advice. Moreover, it is much more difficult to blind interventions at a policy level, because it will usually be obvious from the nature of the intervention that one has received it, thus leaving no room for the idea of a placebo.[6] The net result is that accurately measuring the effect size of a policy intervention in an experimental context is a much more difficult task than measuring the effect size of a drug RCT.

Third, where the underlying reality to be investigated is the physiology of the human body, it is plausible to think that there will be a greater role for translation of research findings from one context to another, than in interventions that are more culturally laden and depend more heavily on shared and changing social meanings. In the latter kinds of case, the fact that we are creatures who interpret our environment, and respond in part to our interpretations, may entail that there are not any universal and unchanging mechanisms of the relevant kind; or that

focus on women's empowerment by 2020 provides a useful case study (Fejerskov 2018; Gates 2019; Wilson 2020).

[5] A point made well by the title of a famous article in the *BMJ*, 'Parachute Use to Prevent Death and Major Trauma Related to Gravitational Challenge: Systematic Review of Randomised Controlled Trials' (Smith and Pell 2003).

[6] Policy trials will often also be cluster randomized trials—where the intervention is not randomized at the level of the individual, but at a higher level entity such as a school or a town. Although the number of people affected by a cluster intervention may be very large, to the extent that the total number of clusters is small, then the power to detect differences may be small, especially if those at the intervention sites interact with those that were determined as controls (Campbell et al. 2012; Hemming et al. 2017).

the mechanisms need to be understood in a broader theoretical context than they otherwise might (Hacking 2007). Researchers or policymakers may be able to devise interventions which work most of the time, but these will be vulnerable to contingencies of social change in ways that pharmaceutical interventions will not.

Even some key proponents of EBM argue that it is a mistake to think that the same approaches can be reused without modification in the realm of policy. As Greenhalgh and Russell put it:

> The Cochrane Collaboration was built on a myth that the judgments required for evidence synthesis are fundamentally technical ones, achieved through the skilled application of tools of the trade such as protocols, data extraction sheets, methodological checklists and evidence hierarchies...In the evaluation of simple clinical interventions (such as drug therapies), this myth approximates reality so closely that it is entirely appropriate to operate as if the world were actually thus. But the world of policy making is not one of transferable and enduring scientific truths, nor is it exclusively (or even predominantly) concerned with what works, and the systematic review movement must adapt accordingly.
>
> (Greenhalgh and Russell 2006: 35)[7]

2.4 Randomization and Internal Validity

Within the experimental research design literature, it is common to make a distinction between internal and external validity. Internal validity is a measure of the quality of the research design: an experiment is internally valid to the extent that it is designed in such a way that it correctly measures the causal effect of an independent variable or variables on one or more dependent variables.

The design of experimental trials is the subject of a massive literature. For our purposes, it will be sufficient to point out a few major factors from the literature on RCTs. In an RCT, eligible patients are allocated at random into either an intervention group or a control group. In testing a new intervention against a control, and allocating participants at random to one of the experimental arms, potential confounding factors both known and unknown are controlled for. Confounding factors such as the placebo effect and treatment bias are ruled out by single or double blinding the trial, so that those giving or receiving the intervention are unaware of whether it is the intervention or the control that is applied. Quite apart from blinding and other controls, there is an elaborate set of requirements

---

[7] Even in the biomedical realm, phenomena such as drug resistance (discussed in Chapter 10.3) show that effective mechanisms will sometimes be fragile and local, rather than unchanging and universal.

for determining sample size (to ensure that, given the expected effect size, the population size is large enough to detect it within set confidence intervals); and ensuring that the endpoints used for analysis are the same as those declared in the trial protocol (to reduce the risk that accidental correlations are misdescribed as causal). Overall, a good summary would be to say that a trial has internal validity (and would meet the methodological threshold standard for publishability) only to the extent that it has been carefully and systematically designed with sufficient care to give a high degree of confidence that the results reported accurately measure the nature and the effect size of the intervention tested.

It is notoriously the case that often the things that really matter are difficult or even impossible to measure accurately. There will often be trade-offs between choosing an end point that is easier to measure, but less directly related to something that really matters (a proxy measure), and one that is more directly related to what really matters, but more difficult to measure. The relationship between the proxy measure and what we really care about may be fairly indirect; and the fact that the intervention has a significant effect size on the proxy measure will not necessarily entail that it also has a significant effect on whatever it is that we actually care about. Moreover, an RCT is designed to measure a mean treatment effect over the trial population. This is compatible with significant variance of effect size within the intervention group. For example a slight mean positive effect overall would be compatible with the intervention being very beneficial for some, but either completely ineffective or slightly harmful for most others.

Randomization is thought to be important or even vital because it provides a very powerful way of reducing bias. However, as Deaton and Cartwright (2018) point out, things are a little more complex than this. What an RCT allows researchers to measure is the mean difference between the control group and the intervention group. As they argue, there will be two sets of differences that combine to form the mean difference between the two groups: first, those differences that are due to the intervention; second, differences that arise in or from differences in causal factors between the two groups.

It is sometimes assumed that randomization will ensure that there are no differences between the two groups, and that the only effect we will see is the genuine effect size, but this is not the case. It all depends on whether the randomization procedure in fact distributes the people between the two groups in such a way that the causal factors that are unrelated to the intervention will cancel out. If we had an infinitely large sample, or if the randomization process was carried out an infinite number of times, and then the results of all those trials evened out, we would expect that all non-interventional differences would even out, but there is no reason to think that this will happen in a single assignment of a finite number of participants.

One way of making this point salient is that if researchers randomize participants into a control and an intervention group, they may end up with results that look 'wrong'—perhaps one group will have a higher percentage of women than the other; or one group will have a lower average age than the other. To the extent that researchers believe that the solution to such 'imbalanced' randomizations is, for example, to re-randomize, this implies that they are in fact relying on a theory about which causal factors are likely to be relevant. Thus, Deaton and Cartwright (2018) argue that one implication is that randomization is *a* method for controlling differences between groups, but it is not the only way or necessarily the best way:

> If we knew enough about the problem to control well, that is what we would (and should) do. Randomization is an alternative when we do not know enough to control, but is generally inferior to good control when we do.
>
> (Deaton and Cartwright 2018: 5)

In short, randomization may seem to give knowledge easily of what will work, without us needing to know anything about the underlying mechanisms and causal structures, but this appearance is misleading. If we do not understand enough about the causal structures to know what needs to be controlled, and the sample size is not large, then the risk of random confounding may be high.

Wainer (2009) gives a useful example of how random patterns can easily lead researchers to infer causes when they are not there. In the late 1990s and early 2000s the Bill & Melinda Gates Foundation funded the conversion of large schools into smaller schools, influenced by the finding that small schools were over-represented in the best performing schools. However, this finding is exactly what would be expected on the null hypothesis—namely that school size made no difference to student attainment. The standard deviation will be much larger in small samples than in large samples, and so small schools will be much more likely to be outliers in terms of the performance of their individual students—both at the good and the bad end—even if school size makes no difference to results. (One easy way of seeing why is to compare it to the results of coin tosses. The chances of getting a sample that is three-quarters heads is vastly higher if you do four tosses than if you do 400, even though the coin is unbiased and has a 0.5 chance of coming up heads each time.) As would be expected from the null hypothesis, when the data were analysed small schools were also over-represented amongst the worst schools, and the shift to smaller schools seemed not to have an independent influence on attainment. The Gates Foundation abandoned its focus on school size after it performed an analysis—having spent over $1 billion.[8]

Given the possibility of random confounding, it is important to be able to estimate the chance that observed results are due merely to random confounding

---

[8] I provide an overview of the Bill & Melinda Gates Foundation's contributions to global health in Wilson (2020).

rather than the intervention. It has been standard to assume that an observed mean difference is statistically significant if there is less than a 5 per cent chance ($p < 0.05$) of it occurring through the random play of chance. This standard for statistical significance has the implication even if everything is functioning exactly as it should, there will be up to one in 20 cases where the mean effect of the intervention on the target variable in the target population is either zero or even harmful, but the random play of chance makes it look as though the intervention was effective.[9] When combined with the fact that researchers are much more likely to write up for publication interventions that 'work', and journals much more likely to publish positive results than negative results, this greatly increases the likelihood that any given publication will report an 'effect' that was in fact due to random distribution. This, when combined with other worries such as p-hacking,[10] and bias famously led Ioannidis (2005) to argue that most published research findings are false.

2.5   External Validity

Even if an RCT provides a correct estimate of the average effect size in the trial population, it may tell us little about whether the same intervention will work in other circumstances. This is because an RCT establishes a conclusion with a high degree of confidence about a particular population. If the population on whom the intervention was tested are not representative of those to whom we now wish to apply the intervention, then the implications of the research for the target population are unclear. To give a common example, if a drug were tested only on otherwise healthy young men between the ages of 21 and 30, it will be far from clear whether the same results should be expected if the drug is given to a frail elderly woman with multiple co-morbidities and who is already taking many other medications (Rothwell 2005).

To sum it up in a slogan, as Cartwright (2013) does, an RCT can show that an intervention worked *somewhere* but not that it will work *here*. A trial is said to have external validity if its results are applicable to a wide variety of other contexts, and in particular if there is reason to think that its intervention will work *here*. To describe a trial as lacking in external validity is not by itself to impugn its

---

[9] As noted earlier, even a mean positive effect is compatible with the intervention having no effect or being harmful for some.

[10] P-hacking is running many statistical tests on a dataset, and reporting only those that allow you to claim statistical significance. If you run hundreds of statistical tests on different variables, then the chances of finding accidental correlations that appear statistically significant are high. This is why, within medicine at least, there are now requirements to upload the trial methodology to a clinical trial register. Detailing the hypothesis that will be tested before the research is undertaken and results obtained makes p-hacking much more difficult. For research on the extent of p-hacking, see Head et al. (2015).

quality as research: it is in the nature of an RCT that the rigour in experimental design that is necessary for internal validity sets limits on how, if at all, the results can be extrapolated to other contexts (Cartwright 2007).

The methods that EBM has encouraged for assessing the quality of evidence—in focusing on RCTs and on systematic reviews and meta-analyses of RCTs—are not very sensitive to the challenge of external validity. As we have seen, the main strength of randomization is supposed to be that it allows a large number of potential confounding factors to be controlled for at a stroke without needing to understand what these confounding factors are. In Fisher's classic formulation,

> Randomization properly carried out... relieves the experimenter from the anxiety of considering and estimating the magnitude of the innumerable causes by which his data may be disturbed.   (Fisher 1935, sec. 20)

However, this point cuts both ways. It also implies something more for challenging for EBM: RCTs do not by themselves tell us anything about *how* the intervention works. The mechanism is treated as a black box; all that a researcher is really entitled to say on the basis of the RCT alone is that changing the value of one variable had an effect size within certain bounds on the value of some other variable or variables.

Without knowledge of how and why a particular intervention has the effect it does, researchers are not in a good position to know whether the effectiveness of the intervention will travel, or if will be purely local. This is one reason why a number of researchers have urged moving away from the idea of fixed evidence hierarchies for public policy, towards the idea of appropriate evidence (Petticrew and Roberts 2003; Parkhurst and Abeysinghe 2016).

We have already mentioned how much more difficult it is to control interventions in public policy than in the context of a pharmaceutical RCT. One of the underlying points was that it is much more difficult to hold constant background causal structures in policy contexts. If something that was presupposed in the background causal structures for the experiment no longer holds in the environment to which the intervention is transferred, we should expect failures of external validity. This is a major reason why it is that policy interventions, even if they have been successfully implemented in one place, will often fail to travel: the intervention was successful only against the backdrop of a set of causal conditions that do not hold in the different context.

Cartwright and Hardie (2012: 66) provide a useful and simple example in their analysis of the California class-size reduction programme. Following a well-conducted RCT in Tennessee, which showed that reducing class sizes was effective in improving reading scores, an attempt was made to try the same intervention in California. However, in California, the intervention failed to improve reading scores. The programme is thought to have failed because some background features of the Tennessee intervention were not replicated in California—such as available classroom space, and enough qualified teachers who could be hired. Making class sizes smaller was part of the team of causes sufficient to improve education in Tennessee, but making class sizes smaller was not sufficient on its own to improve children's education. A rather richer understanding of the background than can be gleaned from an RCT alone will be required for successful public policy interventions (Cartwright and Hardie 2012: 24).[11]

In addition to these points about causal efficacy, it is important to note that a much greater degree of disagreement and contestation is to be expected in the policy realm than in the clinical realm. It is much more difficult to specify what it would be for a policy to 'work' than it is for a drug to work. This point is obscured by some of the examples Chalmers (2003) focuses on, as when he makes much of the bad advice given by Dr Spock in his baby and childcare book, which led to people getting babies to sleep on their fronts, thus increasing the risk of sudden infant death syndrome (SIDS). As Chalmers presents the case, there is only one plausible value in play: it is initially hard to think of a reason why a parent wouldn't want to get their child to sleep on its back rather than its front, given that doing so has no costs, and reduces the risk of death. However, few things in health policy are this simple: there will nearly always be competing values in play, and there will nearly always be some people who object to any given policy, and who would prefer another (Hammersley 2005).

Jonas and Haretuku (2016) discuss how the provision of safe-sleeping advice to reduce SIDS played out in New Zealand. Within Māori and Polynesian communities, there has been a long history of parents bed-sharing with their infants. Bed-sharing 'is seen as positive and beneficial, promoting bonding between mother and child and enabling mothers to comfort and care for their child' (Jonas and Haretuku 2016: 213). The advice given counselled both to put babies to sleep on their backs and against bed-sharing. While the campaign succeeded in reducing SIDS deaths overall,

> the tenor of the anti-bed-sharing message alienated many, particularly indigenous Māori, consequently turning whānau (wider family networks) away from SIDS prevention messages altogether. Some interpreted the campaign as blaming Māori for infant deaths...Several years after the ongoing SIDS prevention campaign was launched, rates of SIDS among Māori remained disproportionately high.   (Jonas and Haretuku 2016: 213)

---

[11] Moreover, in policy contexts there will nearly always be other ways of bringing about the same effect. In this particular case, improved educational resources could have been provided to the children instead. Thinking about different potential routes to a similar result will be important where different potential policies have differing levels of cost-effectiveness, or interfere with liberty to different degrees—as Part II discusses.

Cases of this kind make clear that even where policymakers think that the relevant biomedical evidence is clear-cut, people may nonetheless find advice based on it unacceptable. If so, the advice may either have little effect, or even backfire—leading to reduced trust in state actors or demonstrations. It may be far from obvious what the policy response should be.[12]

In other cases, the complexity of the causal networks and the value-laden nature of the policy choices will require even deeper thought. For example, the distribution of incomes in society is an important social determinant of health; but clearly we also have important reasons to care about the distribution of incomes that are not reducible to the effect that such distributions have on health. It follows that shaping our policy on income tax solely in virtue of its effects on health would be in effect to make the assumption that health is the only thing that matters for social policy, or that health matters more than any other consideration. So even if we were sure that a particular distribution of incomes was necessary to bring about a certain desired pattern of health outcomes in a population, it would be a further question whether doing so would be best, all things considered (Wilson 2009b). What counts as policy success cannot be explained without ethical reflection on how to reconcile various goals and other desiderata.

2.6 Conclusion, and a Way Forward

The idea that health policy should be based on sound evidence seems obvious and uncontroversial, but as we begin to specify how the demand should be interpreted, and what it would entail in practice, things become less straightforward. Despite the initial hopes of EBM, evidential quality cannot be graded and ranked purely in the abstract. While quality of study design matters deeply, it is mistaken to think the mere fact that a report is an RCT, or a systematic review of RCTs, means that it will provide a greater degree of actionable insights for a policymaker than a theoretical, mechanistic, or qualitative study.

Randomized studies have important weaknesses as well as strengths. While randomization is sometimes thought to allow us to control for different factors much more effectively than other experimental methods, this appearance may be illusory. In addition, different causal mechanisms may operate in the context in which the policymakers need to act than where the study was done. Both these limitations of randomized trials—which also affect systematic reviews of randomized trials—imply that what is required for policymaking is not just RCT evidence, but a theory that allows us to understand why the intervention worked

---

[12] Jonas (2016) provides an analysis of the ethical questions about advice-giving that are raised by the case. Chapter 10.4 takes up questions of public trust in the context of vaccination.

where it did, and what modifications would be required to get a similar intervention here.

Much of the rest of this book aims to plot a course through some of the challenges thrown up by the failure of simple models of evidence-based policy. In an old but still useful article, Lindblom (1959) contrasts two diametrically opposed models of policymaking: a rationalist method, and what he describes as 'muddling through'. The rationalist method attempts to combine evidence and values in a scientifically rigorous way across the decision-making process. Decision-makers first work out what all the values are that they are aiming to promote. Then they decide how they should trade-off each value against others (for instance, how much liberty or security is worth how much health?). Next, they collate all the different potential policies that would be relevant to their goals. Finally, they calculate which policy will best maximize the values they had isolated at the first two steps (Lindblom 1959: 79).

The second model, which Lindblom favours and which he takes to be the way that decisions are de facto taken within governments (even if their official approach may say something different), is one of 'muddling through'. The decision-maker who muddles through does not attempt to be exhaustive in the range of policies they consider, but instead focuses only on a few that are salient, and then compares them. This comparison does not usually involve the application of a precise body of theory to guide the comparison, but rather would 'rely heavily on the record of past experience with small policy steps to predict the consequences of similar steps extended into the future' (Lindblom 1959: 79). On this approach, the decision-maker takes a fairly simple goal as given (such as decreasing inequalities in health between rich and poor, while increasing average life expectancy), and simply ignores a large number of other social values which do not seem to directly bear on this goal. Even where the decision-maker does isolate values that are relevant to the chosen goal, muddling through does not attempt to come up with an abstract ranking, but a decision for the specific context.

The rationalist approach might initially sound as if it will be very much the more effective, but Lindblom argues plausibly that it is in fact not possible to use it to make public policy decisions, because we lack the relevant empirical information, and also we lack a theory of how to trade-off the values against one another. The modesty of muddling through seems attractive when compared to the megalomaniac vision of a decision-maker who has a full understanding of all the relevant values, all the interactions between those values, all the possible policies, and all the expected utilities of each of those potential policies in terms of the weighted sum of the selected values.

However, limiting the ambition of social policy to muddling through would thereby also limit our ambition for those whom current systems are failing. Muddling through may make sense from the perspective of the administrator who does not want to make any major mistakes, or to be seen to be rocking the boat—but will seem rather less plausible to someone whose children's lives are blighted by pollution, or who is calling for action in the face of large-scale health problems like obesity or climate change. Structural injustices require rather more than marginal adjustments.

The next two chapters plot a new course between the two extremes of rationalism and muddling through. Chapter 3 deepens and expands the understanding of the problem of external validity, arguing that it applies not just to empirical research, but also to ethical reasoning itself. It argues that some core techniques that philosophers use for ethical reasoning, such as thought experiments—even if internally valid—will often fail to have external validity.

Chapter 4 argues that the combination of failures of external validity in empirical research and in ethics requires philosophers to significantly rethink ethics (and health policy) in line with the insights of complex systems thinking. Large institutions, such as health systems, are not merely complicated, but also complex systems: it is not just that they are comprised of many parts, but also that they are comprised of parts whose interactions make a large difference to how the system as a whole operates. Intervening successfully in such systems requires not so much a cautious 'muddling through', but a rigorous and potentially radical approach that is based in an understanding of the relevant systems structure, and the values of those who partly constitute these systems.

3

Internal and External Validity in Ethical Reasoning

3.1 Introduction

Making wise ethical choices is difficult. One crucial task of ethics is to make such decision-making easier, but without falsifying, or over-simplifying the nature of the choices to be made. The difficult question is what counts as good simplification, and what counts as over-simplification. The main lesson of this chapter is that philosophers have tended to draw the line in the wrong place: they have tended to assume that means such as distinguishing and categorizing cases, discussing thought experiments, and proposing moral principles take us rather closer to wise judgements in real-life cases than they in fact do.

The ideas of internal and external validity, introduced in Chapter 2, are central for this analysis. As we shall see, much recent philosophical work in ethics has assumed that the royal road to better and more rigorous ethical thinking goes via abstraction and simplification, and especially through the creation and analysis of thought experiments.

Such work performs some moves analogous to those that are made within experimental science when researchers design an experiment for internal validity. Just as in the case of experimental interventions, work that focuses on the analysis of abstractions and simplifications will be plausible only on the assumption that there are no serious problems of external validity. If there are systematic problems in establishing external validity in normative ethics, there will be no reason to think that the examination of simpler and more abstract cases will automatically lead to insights that will be helpful in responding to messier and more complex ones.

A deep question about the nature of ethics, and how to do ethics well, lurks in the background of this chapter, before coming fully into focus in Chapter 4. This question is whether ethics should be viewed fundamentally as a theoretical pursuit that aims at the discovery of normative truths, or if it is better to see it as a practical endeavour that attempts to help solve various problems of coordination necessary for sustained and complex human life. My view is that the latter view is correct, and the former seriously misguided. The argument for this conclusion takes place over the next two chapters.

This chapter indicates and interrogates a set of problems about moving between abstractions or simplifications and wise choices in real-life situations, which we will need to contend with regardless of whether we view ethics as fundamentally theoretical or fundamentally practical. As much of the rest of the chapter explores, getting individual real-world cases right is likely to require the kind of messy, complex, and contextual work for which more abstract levels of theorizing can only provide limited guidance. So, even if one's orientation in ethics were entirely towards the theoretical, questions about how to move between judgements based on simple cases and judgements in complex cases would be of central importance. Chapter 4 argues that theoretical approaches fundamentally misconceive the nature of ethical problems in public policy—and that we fail to understand them unless we see them for what they are, as arising and needing to be resolved within complex systems that are constituted by the perspectives and cultures of human beings.

3.2 The Linear Model in Healthcare Research

Questions about the relationship between theoretical advances and practical benefit are particularly urgent in healthcare research, owing to the ubiquity of death and suffering, the technical difficulty and expense of bringing new drugs to market, and governments' own sizeable investments in funding both healthcare research and the successful treatments that result from it. So it is perhaps unsurprising that debate on the relationship between the theoretical and the practical is louder, older, and more voluble in healthcare research than it is in ethics, and contains a rich potential for cross-fertilization.

Approaching the question of the relationship between theory and practice in ethics via debates about translational research in healthcare allows us reframe the dispute between fundamentally theoretical and fundamentally practical approaches to ethics. By starting from outside ethics, in a domain in which it is relatively uncontroversial that there are theoretical facts that inquiry aims to discover, it becomes apparent that even in this domain, it follows neither that more abstract theoretical truths are more worthy of pursuit than more concrete ones, nor that inquiry should be guided by the pursuit of theoretical truths alone.[1]

---

[1] Philip Kitcher has used similar thoughts to argue for an approach to priority-setting in science that he calls 'well-ordered science' (Kitcher 2011a). Briefly, the argument is that there are an infinity of possible theoretical problems that researchers could be working on at any one time, and however many problems they solve, there will always be an infinity remaining. Given this, it makes sense to think that there should be some kinds of norms to guide enquiry, above and beyond merely seeking truths or well-justified beliefs. Where there are an infinity of truths to be discovered, it no longer seems adequate to simply say 'I'm seeking to discover truths.' All human beings could do that all of every day until the end of time, and it would still remain the case that there were an infinity of

The Second World War saw an unprecedented mobilization of science for the war effort. The USA set up the Office of Scientific Research and Development (OSRD) in 1941, whose goal was to 'initiate and support scientific research on the mechanisms and devices of warfare with the objective of creating, developing, and improving instrumentalities, methods, and materials required for national defense' (Roosevelt 1941), leading most notably to the Manhattan Project. As the war drew to a close, the US government faced a choice about the future of science funding: should it continue to funnel scientific research budgets towards short-term solutions for centrally chosen goals, or would it be better to let scientists pursue their own endeavours without regard to practical application?

Vannevar Bush, the Director of OSRD, was tasked with answering this question. The resulting report, *Science: The Endless Frontier*, set the tone for much of the approach to science funding over the next 30 years. Bush began from a distinction between basic and applied research. Basic research 'is performed without thought of practical ends', and 'results in general knowledge and an understanding of nature and its laws' (Bush 1945: ch. 3.3). It provides knowledge that is relevant to answering practical problems, but is not likely to be specific enough to solve particular practical problems. Applied research aims to provide answers to practical problems building on or specifying basic science.

Bush argued forcefully that it is the role of governments to support basic research, and that basic research is best undertaken in universities. Applied research, he argued, was better undertaken by businesses. It was crucial to his account that investing in basic science will (although in ways we cannot yet predict) have significant payoffs in the future:

> One of the peculiarities of basic science is the variety of paths which lead to productive advance. Many of the most important discoveries have come as a result of experiments undertaken with very different purposes in mind. Statistically it is certain that important and highly useful discoveries will result from some fraction of the undertakings in basic science; but the results of any one particular investigation cannot be predicted with accuracy.
>
> (Bush 1945: ch. 3.3)

The idea that theoretical work produces practical benefits—but in unexpected ways that it would be counterproductive to attempt to second-guess—is also at the heart of certain defences of philosophy that emphasize its long-term impact, but deny that this impact could usefully be measured or optimized in the short

---

unknown truths. As Dennett (2006) explores, it is thus quite possible for large amounts of intellectual labour to be expended on problems that, while genuine, are not good uses of the limited resources that we have to devote to inquiry.

term.[2] Of course, one obvious disanalogy is that, when compared to science, there is a decided lack of concerted action in either the commercial sector or in civil society to translate the results of basic normative or other philosophical work into commercializable or socially beneficial results (though the use of game theory in management consultancy and elsewhere might be a possible analogue).

Bush's approach to innovation was amplified and extended by a number of other figures over the next 20 years (as summarized, for example, by Godin 2006 and Balconi et al. 2010), and the resulting approach to thinking about innovation has come to be known as the linear model. The linear model, at its most radical, makes the following assumptions about how innovative products such as new drugs come to be invented:

> (1.1) A clear distinction can be drawn between basic (scientific) and applied (technological and industrial) research...(1.2) Basic or fundamental or prior scientific research is the main or rather the unique source of technical innovation...(1.3) New knowledge acquired through basic research trickles down, almost automatically, to applied research, technology and innovations, even within short time spans.   (Balconi et al. 2010: 5)

Each of these elements of the linear model has been criticized. First, as Stokes (1997) argues, scientific research projects cannot be divided neatly into those that aim to discover fundamental truths about the ways things are, and those that aim at practical application. Some forms of research aim both at practical applicability and at discovering fundamental laws. Louis Pasteur's work, which aimed to address practical questions such as how to avoid milk and beer spoiling, but answered these questions by making fundamental discoveries such as the germ theory of disease, is a key exemplar of this (Stokes 1997: 12–13).

The idea that advances in basic science are always necessary for improved technologies—in Bush's words, that 'basic research is the pacemaker of technological progress' (1945: ch. 3.3) has also been challenged. While there are obvious cases in which new therapies have been developed out of a bedrock of advances in basic science (such as stem cell therapies), it is implausible to claim that healthcare innovation always starts from advances in basic science. Indeed, influential innovation scholars such as Kline and Rosenberg (1986: 288) argue that innovation led by basic research is the exception, rather than the rule.

Third, the idea that new basic science automatically leads to changes in medical practice that benefit patients is also highly questionable. The history of medical science is littered with examples of failures to properly connect basic and applied science. Even a case such as the discovery of penicillin, which might

---

[2] See, for example, Bovens and Cartwright (2010).

initially seem to be an obvious success story, shows how contingent the take-up of basic science into applied science can be.[3]

As a result of the shortcomings of the linear model, a consensus has grown within healthcare research that funders and researchers need to think in systemic terms about how the different stages of research—from the most basic science to the most applied—can best fit together. This approach to thinking has come to be known as translational research. As translational research developed, it became increasingly apparent that optimizing the benefits to patients from research spending requires a rigorous and systematic focus on the pathways and transitions between the different stages of healthcare research, looking for bottlenecks and other opportunities to improve research efficiency.

3.3 Moral Philosophy and the Linear Model

Worries about the lack of direct connection between progress in basic theory and progress in answering more concrete questions apply even more strongly in ethics than in healthcare research. It is plausible to think that much high-quality work in abstract normative theory has very little practical impact, and as we shall see, searching questions can also be asked about current efficiency in moving between ethical questions at different points of the continuum between the fully abstract and fully concrete.

In order to be able to talk sensibly about efficiency in translation between basic and applied contexts in a domain, we need at least a rough model of the continuum of different levels at which research questions can be asked and answered in that domain, and different potential ways of moving between them. Once we have such a model it is much easier to determine whether research insights are currently moving *between* these different levels as effectively as they could be.

Philosophers do not often ask questions about research efficiency in their discipline, or explicitly address the question of how best to move between different levels of abstraction in ethical thinking, and so again the healthcare research literature will provide a starting-point. As Table 3.1 indicates, the journey from the most basic science to interventions that actually benefit patients can be divided into five stages.[4] If an intervention that benefits patients is to be derived from

---

[3] Fleming discovered the antibiotic properties of penicillin by accident in 1928 in the course of his basic research. He published the discovery, and even used it to treat a colleague's conjunctivitis, but was unable to interest his colleagues in it (Fleming 1980; Le Fanu 2000). Finding little take-up for the idea of penicillin's therapeutic use, and having failed to find a way of concentrating it, he abandoned the project (Ligon 2004). It was only ten years later that Florey and Chain chanced on Fleming's paper, and undertook an animal trial, which showed penicillin to be effective in mice, and the potential benefit for human life became obvious.

[4] The translational research literature contains a number of different conceptualizations of this continuum. For an overview of this debate, and a nuanced response to it, see Drolet and Lorenzi (2011).

Table 3.1 Translating normative ethics into actionable results

Stage	Healthcare research stage	Ethics equivalent
1	Basic science.	Discussion of normative theory without any attempt to think about the applicability of moral theory to real-life cases.
2	Proof of concept, e.g. phase 0 trial, testing pharmacodynamics and pharmacokinetics of drug to see that it could in principle be an effective therapy.	Working out what ought to be done in thought experiments.
3	Proof of efficacy, e.g. phase 2 and 3 trials, demonstrating that the drug has a clinical effect under idealized conditions.	Working out what ought to be done in simplified but somewhat realistic cases.
4	Proof of effectiveness, e.g. pragmatic trial, to determine the effectiveness of the drug in real-world clinical settings.	Working out what we should do, all things considered, in real-world situations.
5	Implementation, e.g. policy changes, supported by healthcare research to benefit patients.	Policy changes or other actions to make the world closer to how it should be.

basic science, it must pass through each of these stages.[5] Given the uncontroversial focus in healthcare research prioritization on the benefits to actual patients that are achieved at stage 5, the model immediately raises questions about research funding priorities if it turns out that prioritizing basic science is a markedly less efficient way of achieving patient benefit than, say, research-led improvement of doctors' prescribing habits.

The translational model begins from the observation that much basic research—for example the discovery that a particular protein has a role in regulating inflammation—does not have obvious or immediate clinical application. It may take significant amounts of work to come up with a potential clinical application. The first main place in which there can be failures of translation is thus in moving from (1) basic science to (2) proof of concept in a potential clinical application. The journey from (2) proof of concept to (3) proof of efficacy, i.e. that the intervention works under idealized and controlled conditions, is long and arduous. There are many sites along this journey at which translation could potentially be improved, including better understanding of the limits of animal models, better research regulation, speedier marketing approval, and so on. Once there is proof of efficacy, the next main stage is proof that the intervention will be safe and

---

[5] Of course (and acknowledging this was key to the shift away from the linear model), innovations that benefit patients can also have rather different trajectories. Healthcare innovations need not travel 'from the bench to the bedside', but may instead travel 'from the bedside to the bench' (Marincola 2003). Other important innovations can come from organizing things better *within* a given stage of research (for example by reducing administrative delays in setting up clinical trials).

more effective than its competitors in real-world settings (4). Mere knowledge that an intervention is effective does not save lives by itself—so we can think of the final hurdle as the journey from proof of effectiveness (4) to patients being benefited by the adoption of the intervention into standard clinical practice (5).

We can construct a roughly parallel trajectory for ethics, tracing the kind of steps that would need to occur if new insights in basic normative theory were to lead to ethical improvements on the ground.[6] As in the case of healthcare research, the idea is not to say that ethical change *inevitably* follows this trajectory or that it should; rather, the model allows us to ask questions about where time and other resources are currently being concentrated along this path, and whether transitions between different elements could be improved.

My impression is that many philosophers working within normative ethics, particularly those who make heavy use of thought experiments, adhere (perhaps unwittingly) to a top-down linear model of this kind. In this context, basic research would be discussion of 'pure' moral theory without any attempt to think about the applicability of moral theory to real-life cases; working out what ought to be done in thought experiments would still be close to basic research, while working out what should be done all things considered in real-world situations would be applied research. Just as in a linear model of scientific innovation, it would be assumed that insights will trickle down from more basic theory to more applied contexts, but it is not thought that insights also need to trickle up from 'applied' to more 'basic' theory.[7]

While this model persists, it is now well past its sell by date. Philosophers did often attempt to move straight from basic normative theory to claims about what should be done in the real world, in what Arras (2016) describes as the early 'heroic' phase of applied ethics. But this now seems methodologically naive.[8] The continued deep and unresolved disagreements about matters of fundamental ethical theory mean that even if it is clear what a particular ethical theory would advise in a given case, knowing this does little to help us decide what to do when other undefeated ethical theories recommend something else. Moreover, it has also become clear that ethical theories lack the specificity to give us a definitive

---

[6] My thoughts on the relationship between theory and practice in ethics have been enriched by a number of conversations over the years with Alan Cribb (Cribb 2010, 2011). Cribb (2010: 207) floated the idea of a translational ethics, but treated it merely as a 'heuristic device' to facilitate reflection on the broader project of 'translating ethical scholarship into policy or practice'. This chapter takes the idea of translational ethics seriously and more literally as a potential path for policy-relevant ethics.

[7] See, for example, G. A. Cohen's claim that the subject matter of political philosophy is the search for fact-free fundamental principles of justice: 'facts are irrelevant in the determination of fundamental principles of justice. Facts of human nature and human society of course (1) make a difference to what justice tells us to do in specific terms; they also (2) tell us how much justice we can get; and they (3) bear on how much we should compromise with justice, but, so I believe, they make no difference to the very nature of justice itself' (Cohen 2009: 285).

[8] On this point, see, for example, Caplan (1983), and the critique of the 'engineering model' of applied ethics therein.

steer on many practical questions: as in the case of basic science, it will often take considerable work to develop a claim in fundamental normative theory to a point where it has clear implications for practice, even in idealized cases.

Top-down approaches should be pursued only if there is reason to think that they have external validity—that is, if there is reason to think that the examination of simpler and more abstract cases will dependably lead to insights that will be helpful in responding to messier and more complex ones. The rest of this chapter demonstrates how and why thought experiments, which play a crucial role within such approaches, frequently fail to have external validity. In the wake of this critique, Chapter 4 builds a more bottom-up and fluid view of ethical reasoning.

3.4 Thought Experiments

Thought experiments as I understand them here are toy ethical cases that are designed to simplify an ethical problem along a number of dimensions, thus making the problem more philosophically tractable.[9] The following classic thought experiment by Rachels (1975) provides a good initial example. Rachels aims to answer the question whether killing is, in itself, worse than letting die. He draws the reader's attention to the fact that he has constructed a pair of cases that 'are exactly alike except that one involves killing whereas the other involves letting someone die':

> In the first, Smith stands to gain a large inheritance if anything should happen to his six-year-old cousin. One evening while the child is taking his bath, Smith sneaks into the bathroom and drowns the child, and then arranges things so that it will look like an accident.
>
> In the second, Jones also stands to gain if anything should happen to his six-year-old cousin. Like Smith, Jones sneaks in planning to drown the child in his bath. However, just as he enters the bathroom Jones sees the child slip and hit his head, and fall face down in the water. Jones is delighted; he stands by, ready to push the child's head back under if it is necessary, but it is not necessary. With only a little thrashing about, the child drowns all by himself, 'accidentally', as Jones watches and does nothing.   (Rachels 1975: 79)

---

[9] To anticipate some of our conclusions, thought experiments fall on a continuum from the more austere to the more richly described. All thought experiments involve selectively presenting some, rather than other, elements of a world for contemplation. This selective presentation of a world is also a feature of literature, opening the possibility that works of literature can be thought experiments.

Rachels reasons that examining two cases that differ in precisely one respect will allow the reader to isolate the ethical difference (if any) that there is between killing and letting die. His language and approach suggests that the use of the thought experiment should be understood in the same light as a scientific controlled experiment—and this is a theme that I shall take forward in what follows.

Rachels's thought experiment belongs to a class of thought experiment that I shall call an *interventional sequence*. In an interventional sequence, a base case is presented, and then modified in one or more further cases with the aim of discovering the effect that the altered feature has on ethical judgements. We can distinguish interventional sequences from *one-off* thought experiments, where a single scenario is presented for ethical consideration without attempting to modify that scenario in a controlled way.

Some thought experiments are advanced as clear or paradigm cases for the appropriateness or inappropriateness of a particular ethical judgement. Peter Singer's Shallow Pond example provides a memorable example: 'if I am walking past a shallow pond and see a child drowning in it, I ought to wade in and pull the child out. This will mean getting my clothes muddy, but this is insignificant, while the death of the child would presumably be a very bad thing' (Singer 1972: 231). Call these *clear cases*. Clear cases can be either one-offs, or form part of interventional sequences.[10]

Other thought experiments are presented as *problem cases*, in which what should be done is thought to be either by its nature difficult or unclear; or where the case creates a problem for consistency with other judgements that are commonly held. Problem cases can also be either one-offs or form part of interventional sequences. Bernard Williams's famous thought experiments critical of utilitarianism were one-off problem cases. For example, George the Chemist, who is offered a way out of long-term unemployment through a job designing chemical and biological weapons, aims to raise a question about whether a moral theory that gives no intrinsic weight to integrity can account adequately for some fundamental features of moral life (Smart and Williams 1973: 97–100). The trolley literature, following Foot (1967) and Thomson (1976), presents a complex interventional sequence of problem cases about the conditions under which it is permissible to cause harm to one person in the course of averting the death of a greater number.[11]

---

[10] Rachels's interventional sequence relies on the assumption that both Smith and Jones present clear cases of morally wrong behaviour.

[11] For the sake of completeness, there can also be mixed interventional sequences—either where an initially clear case is modified progressively in order to shed light on what would otherwise be thought to be a problem case (for example, by piecemeal adding and analysing additional factors to a clear case), or where an initially problem case is modified progressively in order to show how it can be thought of as a clear case. Unger (1996) provides an example of such a mixed approach.

3.5 Internal and External Validity

What kinds of support can thought experiments in normative ethics provide for normative claims? What is it for them to be rigorous? The view that the terminology of thought experiment suggests, and which is supported by many practitioners of thought experiments in normative ethics, is that a thought experiment is a kind of experiment. Just as scientific experiments are rigorously controlled in order to answer precise research questions, while minimizing the risk of contamination of the results by confounding factors, so should thought experiments be.[12]

If thought experiments are a kind of experiment, and depend for the epistemic force of their results on the rigour of their research design, it is worth beginning any search for rigour in thought experiments with the very much better established literature on experimental research design.[13] As we saw in Chapter 2, within this literature, it is common to make a distinction between *internal* and *external* validity. Internal validity is a measure of the quality of the research design: an experiment is internally valid to the extent that it is designed in such a way that it correctly measures the causal effect of an independent variable or variables on one or more dependent variables. Even if a clinical experiment is internally valid, it may tell us little about whether the same intervention will work in other circumstances. An experiment such as a randomized controlled trial (RCT) can show that an intervention worked *somewhere* but not that it will work *here* (Cartwright 2013). A trial is said to have *external validity* if its results are applicable to a wide variety of other contexts, and in particular if there is reason to think that its intervention will work *here*.[14]

---

[12] Kamm puts it thus: 'Real-life cases often do not contain the relevant—or solely the relevant—characteristics to help in our search for principles. If our aim is to discover the relative weight of, say, two factors, we should consider cases that involve only these two factors, perhaps artificially, rather than distract ourselves with other factors and options' (Kamm 1993: 7).

[13] Someone might wish to deny that thought experiments are a genuine kind of experiment, or might claim that while thought experiments are a kind of experiment, rigour in thought experiments is *sui generis*, and thus there is nothing to be learned from experimental research design. Claiming that thought experiments are not supposed to be rigorous in the way that scientific experiments are does not by itself provide support to the claim that philosophical thought experiments *are* rigorous. Moreover, the problems of internal and external validity that will be uncovered in the next sections are not dependent on the claim that thought experiments are scientific experiments, but derive from the fact that thought experiments in normative ethics provide a simplified model of a kind of scenario that might happen, and use this model as a way of reflecting on and gaining understanding of the actual world. The challenge of rigorously moving between the world of the thought experiment, and the actual world is inextricable from the use of thought experiments in ethics, and cannot be dispelled by a disavowal of scientific methodologies.

[14] Internal validity is a necessary condition for external validity. The distinction between internal and external validity could thus be thought of as analogous to the distinction between validity and soundness in arguments. An experiment that leads to results that happen to be widely applicable, but does so despite a lack of rigour in its research design, would be equivalent to an invalid argument with a true conclusion.

Making the distinction between internal and external validity allows us to see that there are *two* crucial questions about rigour in thought experiments. The first, on which the next section focuses, is the nature of internal validity in ethical thought experiments. As a first approximation, we might say that, by analogy to a clinical experiment, a thought experiment is internally valid to the extent that it allows its readers to make judgements that are confident and free of bias or other confounding factors about the hypothesis or point of principle that it aims to test. However, as we shall see, the very idea of internal validity in thought experiments is significantly complicated by the fact that thought experiments are a type of fiction, and that making judgements about cases presented in thought experiments is to respond to fictions.

The second crucial question, which is the subject of Section 3.8, is external validity. To what extent do ethical judgements that are correct of the world of the thought experiment generalize to a wide variety of other contexts, including ethical decision-making in the actual world? To anticipate the results of this analysis: just as in the case of clinical trials, designing for internal validity in thought experiments is vital. But just as in clinical trials, there is no easy route from internal to external validity.

3.6 Internal Validity in Thought Experiments

I distinguished between four types of thought experiments: (1) one-off clear cases, (2) one-off problem cases, (3) interventional sequences of clear cases, and (4) interventional sequences of problem cases. One-off clear cases are introduced standardly for rhetorical purposes as the uncontroversial starting-point from which more contentious results will be derived. There is usually little attempt to establish that the judgements about the clear case, which it is assumed that the audience also shares, are in fact correct. Of course, there is ample room for scepticism even about the use of one-off clear cases—the author might be mistaken that something is a clear case, or about what it is a clear case of, for instance. But given the rhetorical role that the one-off clear case plays, little of philosophical importance turns on the precise way in which it is presented. To the extent that there is a clear case to be captured, and an author's thought experiment currently fails to do this, the author's thought experiment can easily be swapped out or fixed up. As the focus of this chapter is on thought experiments as a way of *improving* moral insights, I shall have little more to say about one-off clear cases.

Problem cases require much more precision in their setup. By their nature, they are likely to be more controversial and more surprising than clear cases, and less likely to lend themselves to purely rhetorical use. The case setup needs to do at least the following three things to have a plausible claim to internal validity.

First, what in the context of a legal analysis would be described as the *facts of the case* need to be described clearly, economically, and consistently. It should be clear, for example, whether the problem is supposed to arise from the perspective of an omniscient narrator, or from the perspective of a fallible and limited protagonist. While adding colour may make the case enjoyable to read, this should not come at the expense of confusing or needlessly complicating the scenario presented. Second, the case needs to do more than to point out the conceptual possibility of a problem; it needs to be able to allow the reader to appreciate the problem as a concrete one occurring in the context of a possible scenario. Third, the problem and analysis needs to be genuine. Insofar as the problem relies on the set of choices available to the actors within the thought experiment being significantly reduced, this needs to be specified and at least justified (where justification might include mere stipulation).

Internal validity in interventional sequences of clear cases is structurally similar to internal validity in controlled clinical trials. In both cases, there needs to be a control case and one or more interventional cases, and the interventional cases must differ from the control case only in the ways described in the methodology. Thus, in both cases, researchers need to be able to specify what research question is being tested; and to specify which variables are being modified between the cases, and which are being kept fixed. For example, if the thought experiment requires comparing two cases which differ only in a single respect, readers need to be confident that the cases *do* only differ in this one respect. In addition, the author will need to control for potential ordering effects in the presentation of cases.[15]

Attaining internal validity in interventional sequences of problem cases will require their authors to combine the lessons of internal validity of both one-off problem cases and interventional sequences. A genuine problem needs to be described clearly, economically, and consistently in enough detail for the reader to be able to see the problem as arising in a particular context. This base case then needs to be modified in a controlled way. It is vital that what is kept the same and what changes between the cases is explored rigorously both at the level of description of the cases, and at the level of moral analysis. This will frequently be challenging: what makes something a problem case is either an intrapersonal clash between intuition about the case, and the deliverances of theory; or an interpersonal clash between different individuals' judgement about the case. Given this level of uncertainty, there may well be disagreements about what is morally relevant in the case as described and so about which features of the case should be kept constant, and which modified in order to make progress.

---

[15] See, for example, Schwitzgebel and Cushman (2015).

A surprisingly large number of classic thought experiments in normative ethics fail to reach these basic requirements of internal validity. For example, here is Nozick introducing the problem of innocent threats:

> If someone picks up a third party and throws him at you down the bottom of a deep well, the third party is innocent and a threat; had he chosen to launch him-self at you in that trajectory he would be an aggressor. Even though the falling person would survive his fall onto you, may you use your ray gun to disintegrate the falling body before it crushes and kills you?   (Nozick 1974: 34)

Nozick's thought experiment is memorable but problematically unclear. The pres-entation of the case conflates the perspective of the omniscient narrator with that of the person at the bottom of the deep well trying to work out how to respond to a fast-falling human body. It is possible to view such a scenario from the perspec-tive of an omniscient narrator who is able to state with certainty all the facts relat-ing to the case, or from the perspective of the time- and information-starved person dealing with the situation, but not both simultaneously.[16] Quite apart from this, Nozick's attempts to enliven the case make it needlessly difficult to interpret. Why is the person at the bottom of the deep well (have they themselves been thrown down, or are they there voluntarily)? If the well is that deep, would the person at the bottom be in a good position to judge in good time that what has been dropped in is a human body? Why introduce the idea of a ray gun, which suggests a science fiction scenario, without further contextualizing this?

In other cases, something that the author thinks is imaginable may not be concretely imaginable, or be imagined very differently by different readers. For example, Kamm (2006: 352) gives the Reach Case, providing only the following description: 'Suppose that I stand in a part of India, but I have very long arms that reach all the way to the other end of India, allowing me to reach a child who is drowning in a pond at a great distance.' Kamm intends that the Reach case will allow the reader to see that it should be treated like a case where the child is near. However, it is far from clear how one is supposed to imagine the case. It seems from the context that Kamm intends that the extra-long arms be an integrated part of the individual's body. This raises a number of challenges for coherently imagining the case: is the rest of the person supposed to be scaled-up in propor-tion to their thousand-mile-long arms? The context suggests not. But if we are supposed to imagine an ordinary-sized human being with arms a thousand miles long, how is their anatomy supposed to work? How much would

---

[16] Given the described physical constraints, it seems unlikely that a person who actually was at the bottom of the deep well would be in a good position to judge (a) if they would definitely be killed by the falling body, or (b) whether the person falling would definitely survive the fall if not disintegrated by the ray gun.

thousand-mile-long arms weigh? How would the weight be supported? Where do they buy their shirts?

As Elster (2011) discusses, readers need to be able to imagine in sufficient detail not only any outlandish elements of the thought experiment such as thousand-mile-long arms, but also the implications of these outlandish elements for ethical norms and practice within the world of the thought experiment. Merely stipulating that the world of the thought experiment is different in various ways from the actual world, but not reconstructing the rest of the world of the thought experiment in the light of these assumptions, is to fail to take seriously the case as a genuine case.

3.7 Reproducibility, Fiction, and Thought Experiments

Some readers have accused this line of objection to the setup of thought experiments of being uncharitable: in context, what Nozick or Kamm intends is clear enough. However, the appeal to charity in interpretation sits a little uncomfortably with the idea that the rigour of a thought experiment comes through its methodology, as we can see by comparison with the case for RCTs.

Experimental research design places the idea of reproducibility at its core. It is a basic requirement of science that, in order for an experiment to be publishable, its methodology must be written up in a way that allows the reader to appraise its internal validity, and to allow a suitably skilled team to reproduce the results. Readers will not usually try to reproduce the results obtained (given constraints of time, resources, and equipment), and only a small percentage of experiments will ever be reproduced, but it is core to the purpose of reporting methods that the results would be such as to be reproducible given the statement of the methods.[17]

If a thought experiment is supposed to be reproducible by the reader, it is vital that the author presents the reader with everything she would need to run the thought experiment herself. Where the author's scenario is indeterminate or under-described, and there are a variety of ways of filling in or imagining the stated scenario, and these different ways could plausibly have different ethical implications, then the author has not done enough to ensure all readers are attending to the same case. In that instance, the case is not reproducible: the scientific equivalent would be leaving some core elements of the methodology indeterminate, so that it was unclear, for example, what dose of a particular drug was given or what the inclusion and exclusion criteria for the trial were.

---

[17] For some challenges for reproducibility, see, for example, Stodden (2014).

Reproducibility in science requires not just that an experiment with the same essential features can be set up on the basis of the described method, but also that the results can be replicated.[18] Thought experiments, as they are used in normative ethics, are intended to give rise to judgements about reasons, rather than merely mechanistically caused responses. So while it is clear that thought experiments require that the case is reproducible, in the sense that it is clearly and consistently described, the degree to which reproducibility of judgements is required is less clear.

Case reproducibility in thought experiments faces a deep, but obvious, problem: thought experiments are a kind of short fiction.[19] Just as in longer fictions, the author presents a world in which certain events occur and choices are made, and as readers we are asked to attend to this world.[20] Just as in longer fictions, only a very small number of the questions that could be asked about this world are answered by the materials with which the author has provided us: the rest are left indeterminate. Knowing this unavoidable ambiguity, writers of literary fiction often choose to thematize it: rather than attempting the impossible task of presenting a world in such detail that all questions a reader might have about it will be answerable, they present a world that is deliberately spare, fragmentary, or filtered through multiple incompatible perspectives. This way, each reader completes the world of the text on the basis of what he or she brings to it.

Writers of thought experiments also unavoidably depict worlds that are underdescribed. But where allowing each reader to complete the world in his or her imagination may be a virtue from the perspective of the literary writer, it is far from clear that it is a virtue if in constructing the thought experiment its author is attempting to conduct, or get the reader to conduct, a controlled experiment. If the methodology of thought experiments consists in getting individuals to interpret and to make judgements about very short fictions that are indeterminate in a large number of aspects, how is reproducibility to be ensured?

This problem of underdetermination is potentially so intractable for the use of thought experiments in ethics that it is usually ignored rather than squarely addressed. I take it that the problem is usually supposed to be ameliorated by a convention of *authoritative authorial ethical framing*. On this convention, the case

---

[18] The discovery of widespread failures of reproducibility of results has led to a perceived crisis in disciplines such as psychology (Pashler and Wagenmakers 2012).

[19] As Elgin explains, 'We perform thought experiments by imagining a scenario in which something happens—a sequence of events with a beginning, middle, and end. Thought experiments can be construed as tightly constrained, highly focused, minimalist fictions, like some of the works of Jorge Luis Borges. If the minimalist stories of Borges are genuine fictions, there seems no reason to deny that thought experiments are too' (Elgin 2014: 230).

[20] See, for example, Kamm: 'This may be just an autobiographical fact, but I don't really have a considered judgement about a case until I have a visual experience of it. I have to deeply imagine myself in a certain situation, with an open mind...What I am saying is that, in order to have a judgement about a case, you really have to situate yourself in the case' (interview in Voorhoeve 2009: 22).

raises the ethical question or questions that the author of the thought experiment says it does; the ethical issue that the case raises is not subject to dispute. To further spell out the implied convention, the author of the thought experiment has, by definition, specified all the elements of the case that are morally relevant. No morally relevant differences other than those that have been stipulated by the framer of the thought experiment apply to the situation. Although each reader will fill out the details of the case in their own way in imagining it, each reader may only add colour and detail that is morally irrelevant.[21]

Thought experiment designers further attempt to finesse the problem through the use of an omniscient authorial voice, able to take in at a glance and to relate events in their essentials. The voice is able to tell us clearly and concisely what each of the actors within the thought experiment is able to do, their psychological states, and intentions. The authorial voice will often stipulate that choices must be made from a short predefined menu, with no ability to alter the terms of the problem. For example, the reader may be presented with two choices: to pull a lever, or not to pull it. The world of the thought experiment may also be stipulated to operate according to laws that are plainly false of the real world. In this case, responses that would be likely to be effective in real-world analogues of the thought experiment may be stipulated not to work, and other responses, which would be unlikely to be effective in real-world analogues, may be stipulated to be effective.

Such techniques face a tension: in so far as the case is very sparsely described and is stipulated to have only the facts or features that its author says, then there is little to imagine and it is difficult for its readers to experience the case as a genuine case. But the more difficult it is to experience the case as a genuine case, the less it is plausible to claim that thinking about the case has added something additional to thinking about the problem in purely abstract terms, and the less confident we should be that the judgements that arise from the case have clear implications about the actual world. Alternatively, the richer and more realistic the case, and the more space the writer gives the reader as interpreter to decide what is salient, the easier it is for the reader to experience the case as genuine; but at the same time the less plausible it is to think of the case as akin to a controlled scientific intervention, and the more scope there is for disagreements in judgements about the case.

Failures of reproducibility of judgements are fatal for purported clear cases. A case fails to be a clear one if, given a perspicuous description of the case, there

---

[21] Applying this convention to Rachels's cases of Smith and Jones, one would be free to imagine Smith being of a variety of different ages, and as having a variety of reasons for wanting the inheritance; but any features attributed to Smith must also be attributed to Jones, and the reader is not free to imagine that the two have a feature that would call into question their moral responsibility for the killing or letting die (neither is a psychopath, or subject to mind control).

is a significant degree of disagreement on what should be said about it. Persistent disagreements about what should be done do not fatally undermine the usefulness of a problem case, but they do raise deep questions about the work that such thought experiments are supposed to do in ethical reasoning. There are at least three possibilities. First, failure of reproducibility of judgements could show that the thought experiment is flawed in its design and description. Second, there may be persistent blameless disagreement even among competent moral agents who have the same understanding of the ethically relevant features of the case— caused, for instance, by the factors that Rawls isolates as the 'burdens of judgment' (1993: 53–7). Third, some might have greater expertise than others in responding to thought experiments. On such a view, if nonexperts fail to reproduce expert judgements about thought experiments, this is no more of a problem than if nonexperts cannot reproduce experimental results that require advanced lab skills and years of specialist training.[22]

Regardless of the conventions that are in play for interpretation of thought experiments, ethical judgements elicited by a thought experiment apply in the first instance to the world of the thought experiment. Just as with other fictions, it is an open question how ethical judgements that can be made correctly about a fictional world bear on the actual world. Obviously, if ethical judgements that were correct of the world of the thought experiment applied only to the world of the thought experiment and had no implications for the actual world, analysis of thought experiments would not be able to contribute to resolving the difficult ethical problems that are the point of departure for ethics in the first place. So, a defence of thought experiments as a methodology in ethical research needs to be able to show not just that we are able to make helpful and accurate ethical judgements about thinly described fictional cases, but that such judgements bear in an enlightening way on ethical judgements about the real world. However, to say the least, it is unclear why it should be the case that we can make wise decisions about complex real-life cases by rigorously analysing cases that are simpler (often radically simpler) than the real-life cases that we are aiming to resolve wisely (Dancy 1985: 166).

---

[22] In response to the worry that some (perhaps many), say that they cannot form a confident judgement or even form a judgement at all about cases that are baroque or distant from everyday experience, Kamm suggests that, even among philosophers, there may be relatively few who have the wherewithal to be able to reproduce her judgements about particularly complex cases: 'Having responses to complex and unfamiliar cases requires that one see a whole complex landscape at once, rather than piecemeal. This often requires deep concentration. Only a few people may be able to respond to a complex case with a firm response...the "Princess and the Pea" is the fairy tale best associated with the method I describe: it tells of someone, despite much interference, who cannot ignore a slight difference in a case that others may never sense' (Kamm 1996: 11).

3.8 The Problem of External Validity

Thought experiments can lack external validity in at least two ways. First, if the ethical judgements that can be established as appropriate in the world of the thought experiment depend on features of the normative context that are not shared in other normative contexts. Call this *normative contextual variance*. Second, if the ethical judgements that can be established as appropriate in the world of the thought experiment presuppose causal structures that are relevantly different from those that are present in other contexts. Call this *non-transferability of causal structures*.

3.8.1 Normative Contextual Variance

Rachels's case contrast strategy presupposes that if there is no moral difference between what Smith does and what Jones does in the precisely equalized cases, then there is no intrinsic moral difference between killing and letting die. More broadly, the underlying thought seems to be that if there has been a proper setup and analysis of the two cases (i.e. if there is internal validity in the design of the thought experiment), then the fact that a given moral feature makes a difference in the thought experiment shows that it makes a difference everywhere. Kagan (1988: 12) describes this as the *ubiquity thesis*. If the ubiquity thesis were true, then external validity would come for free with internal validity.

However, it is at best unclear whether the ubiquity thesis should be accepted. It is important to notice how strong the thesis is. First, the thesis requires that if we can find *any* pair of precisely equalized cases, and show that the feature which is different between the two cases either matters, or does not matter, then this result applies to all other contexts. But ethical principles and considerations are often claimed to interact with one another in holistic ways. On this view, there are scenarios in which ethical considerations that favour acting in certain ways in many or most cases no longer provide a reason in favour of acting in that way, and may even change polarity and provide a reason against acting in that way. For example, in usual circumstances, the fact that doing X will provide someone else with pleasure speaks in favour of it, but there are readily imaginable scenarios where the fact that something will create pleasure for someone would count against it (suppose the pleasure is sadistic).

Many moral philosophers, and many nonphilosophers thus endorse what Kamm describes as the Principle of Contextual Interaction, namely that a moral property can 'behave differently in one context than in another'. If the Principle of Contextual Interaction is correct, then it might be the case that 'in some equalized contexts, a harming and a not-aiding will be judged as being morally equivalent, yet in other equalized contexts, they will not be' (Kamm 2006: 17).

The ubiquity thesis licences inferences of a universalistic kind that, from the perspective of the Principle of Contextual Interaction, look to be obviously unsafe. For example, the ubiquity thesis licences the inference that if we can find one pair of cases (such as Smith and Jones) which are precisely equalized and differ in only one feature, and that there is no moral difference between these cases, then *for any pair of cases* that are precisely equalized and differ only in this one respect, there will be no moral difference between the cases.[23] The fact that the ubiquity assumption is itself controversial and has been subjected to many purported counterexamples, suggests that it would make work in normative ethics less rather than more rigorous if writers in normative ethics were to pre-suppose its truth in using thought experiments. Failures of external validity as a result of normative contextual interaction must be expected and planned for if the use of thought experiments in ethics is to be responsible.

3.8.2 Non-Transferability of Causal Structures

In the case of scientific experiments, failures of external validity arise from a non-transferability of causal structures: something that was presupposed in the background causal structures for the experiment no longer holds in the environment to which the intervention is transferred. As we saw in Chapter 2, this is a particular problem for interventions at a policy level, where there are often indefinitely large numbers of background features presupposed by successful randomized controlled interventions.

If we take thought experiments in any way seriously as a basis for possible action, we will have to contend with exactly the same problem. This problem arises particularly acutely in cases where the thought experiment purports to model a real-world choice context, but has a significantly different experiential, psychological, causal, or epistemic structure from the real-world context (Bauman et al. 2014). Thought experiments about risk are particularly problematic in this regard. In the world of the thought experiment, it will often be stipulated that there is certainty of effect: each of the defined choices will bring about its stipu-lated effect with certainty. It will be possible to identify in advance who will bene-fit, and who will lose, from each of the predefined choices. Even where the choice is stipulated to bring about the desired effect a certain percentage of the time,

---

[23] This leaves such attempts to establish the equivalence of harming and not-aiding under signifi-cant threat of counterexample: 'if we can find even one set of comparable cases in which a harming is morally worse than a not-aiding, we rebut the Equivalence Thesis, for while a single positive instance cannot prove a universal claim, a single negative instance can defeat it' (Kamm 2006: 17).

these stipulated probabilities are entirely accurate. Barely any real-world cases of risk imposition have this structure.[24]

Here is an example from an influential recent article by Frick, which in fact begins by acknowledging that contractualist accounts of harm have so far focused problematically on '*certain* harms to *known* individuals' (2015: 178), where such cases are in fact rare. Frick aims to extend contractualist analysis to the types of risk that occur in social policy, centring his analysis around the following case:

> Mass Vaccination (Known Victims): One million young children are threatened by a terrible virus, which is certain to kill all of them if we do nothing. We must choose between mass producing one of two vaccines (capacity constraints prevent us from producing both):
>
> Vaccine 1 is certain to save every child's life. However, the vaccine will not provide complete protection against the virus. If a child receives Vaccine 1, the virus is certain to paralyze one of the child's legs, so that he or she will walk on crutches for the rest of his or her life....
>
> Vaccine 3 is sure to allow 999,000 children to survive the virus completely unharmed. However, because of a known particularity in their genotype, Vaccine 3 is certain to be completely ineffective for 1,000 identified children. These doomed children are sure to be killed by the virus if we choose Vaccine 3.
>
> (Frick 2015: 181–3)

As a moment's consideration reveals, the case described differs in various salient ways from the causal mechanisms by which communicable diseases are spread in the actual world. First, in the actual world, exposure to the infectious disease pathogen is a necessary, but not a sufficient condition for developing a clinically significant infection: whether a clinically significant infection follows depends on the interaction of features of the host, environment, and pathogen. Second, in any real-life scenario there would be a distribution of degrees of severity of clinical symptoms, rather than a sharp divide into two homogeneous groups. Moreover, in the real world, vaccines are rarely 100 per cent effective: they are likely to be more effective for some than for others; some may be allergic to the vaccine, or unable to attend on the day of the mass vaccination. So, in any real-world scenario, whatever is done, some children will not be vaccinated. Thus, in any case that has the causal structure of an infectious disease outbreak in the *actual* world, the number of persons exposed to the pathogen, and the number of persons who

---

[24] As Fried (2012a, 2012b) argues, there is an additional worry about external validity and contextual variance here. Individuals' judgements about cases involving *certain* harm are often difficult to reconcile with their judgements about *risks* of harm. It is unclear why it should be assumed to be more methodologically robust to examine the ethics of risk imposition in a way that requires individuals to imagine away the everyday experiences they have of risk, than by building on these experiences.

will develop clinically relevant symptoms will not be knowable in advance, but will only be able to be estimated (often within wide confidence intervals) on the basis of mathematical models.

Third, the case is stipulated to be one of a deadly virus, but no discussion is given of the mode of transmission. In any usual case of a virus that attacks human beings, and can be transmitted from one human being to another, herd immunity will be relevant. That is to say, the likelihood of being exposed to the pathogen depends on the degree to which others have been vaccinated (and so your likelihood of getting the disease if you yourself are not vaccinated is much lower in an environment where all others are vaccinated than where none are).[25] Frick's Vaccine 3 scenario elides this fact, assuming that if the vaccine is ineffective for 1,000 children then these children will be (a) sure to be exposed to the pathogen, and (b) die as a result of being exposed to the pathogen.

Fourth, it is unclear why the children for whom Vaccine 3 would be ineffective could not be isolated until the disease passes (they are stipulated to be identifiable, and to be only 1,000 out of 1 million). Overall, there are so many differences in underlying causal structure between Frick's case and any actual vaccination policy case, that even assuming internal validity for Frick's analysis of the thought experiment and leaving on one side any questions of normative contextual variance, the thought experiment has very limited relevance for actual vaccination policy decisions.

3.9 Conclusion

Philosophers have often thought that abstraction and, in particular, thought experiments are a key way through which the rigour of ethical theorizing is increased. However, theorists in ethics have thus far failed to distinguish adequately between two kinds of rigour: internal and external validity. External validity is both more important and more difficult to ensure, but philosophers have tended overwhelmingly to focus on internal validity.

The results of our analysis place thought experiments in ethics in a significantly weaker position than randomized clinical trials with respect to internal validity, but roughly on a par when it comes to external validity. Even if a thought experiment *is* internally valid, this does not provide a strong reason to think that it is externally valid—due both to the possibility of normative contextual interaction, and to the likelihood that the actual world has a relevantly different causal structure from the world of the thought experiment.

---

[25] An exception to this general principle of vaccination would be tetanus, which is not communicable from person to person. However, given the description of the case, there does not seem to be any reason to suppose that Frick has in mind a noncommunicable virus.

Some readers have wrongly interpreted this as an argument for the abandon-
ment of thought experiments and idealization in ethics. Some of these readers
have leapt to the defence of thought experiments, arguing that it is not possible to
do ethics without considering hypothetical scenarios. This purported defence of
thought experiments against my argument misfires. I have not argued against the
use of thought experiments *tout court*, but for a much greater reflexivity in how
they are used. Just as the failure of internal validity to guarantee external validity
does not imply that randomized clinical trials cannot be externally valid, so the
failure of internal validity to ensure external validity in thought experiments does
not imply that thought experiments cannot be externally valid. In both cases, the
relevant question is the extent to which a judgement that appears correct in one
context reliably transfers to other contexts.

Debate in philosophy often proceeds through the search for counterexamples,
which purport to show a particular theory cannot be correct, because it cannot
handle (or gives the 'wrong' result) in one or more hypothetical cases. My orientation is different. The best question to ask of a theoretical construct is not
whether there are conceivable circumstances in which it would give wrong or
problematic results (of course there will be), but to specify the range of circumstances in which it provides results that are useful and reliable. A simple thought
experiment can be useful if it alerts us to a conceptual or regulatory possibility
that would otherwise be overlooked, just as a very simple two equation model
such as the Lotka–Volterra predator–prey model, can be useful for understanding
aspects of the dynamics of predation. However, in neither case does usefulness in
one context imply that the thought experiment or model would by itself provide a
wise basis for making policy.

The next chapter argues that the scale of the challenge posed by external validity both in causal and interventional reasoning, and in ethical reasoning, has radical implications for the philosophy and ethics of public policy. The default should
no longer be to start from simplistic causal models or thought experiments, while
being aware that these approaches will exclude some features that would be relevant for real-world decision making. Rather, both practitioners and philosophers
should start from the premise that social processes are complex systems. Doing so
requires that a paradigm shift is necessary—towards a more experimentalist- and
pragmatist-inspired conception of philosophy.

4

Ethics for Complex Systems

4.1 Introduction

Public policy is an arena in which there are likely to be widespread failures of external validity. Even if there is a body of well-conducted research, which shows that a policy intervention worked somewhere, this is not enough to establish that the intervention will work here. Chapter 3 showed that similar concerns about external validity arise within normative ethics itself. A significant body of research from thought experiments may do little to provide ethical insight into the concrete problems we face here and now. Thus, public policy interventions must face challenges of external validity not only in the realm of facts, but in the realm of values, and in how the two interrelate.

This chapter argues that the scale of the challenge requires a similarly sizeable response, changing not only the way that practitioners approach evidence collection and interventions in policy, but also the way philosophers conceive of the task of philosophy. The default should no longer be to start from simplistic causal models or thought experiments, while being dimly aware that these may be inadequate. Rather, researchers should start from the premise that social processes are complex systems.

Complex systems approaches start from a very different set of assumptions than those that are implicit within common-sense accounts of causation, or thought experiments. We'll see this in more detail shortly but it will be helpful to begin with an intuitive comparison. Much common-sense reasoning about causation builds from a base model of something like a billiard table—in which everything is static until a force is applied from the cue to the cue ball, and this then strikes another ball and causes it to sink into the pocket. Similarly, the thought experiment is designed in a way that allows everything to be screened out and kept static, apart from the one ethical question of interest.

Biological and social systems behave very differently from these static models. Where they exhibit stability, it is a *dynamic stability*. Key variables of interest do not stay the same because nothing is changing, but rather because they are kept within boundaries by the dynamic interactions of a number of other regulatory systems. Homeostasis, the process by which living creatures maintain their critical internal states within a fairly narrow range of values despite changes to the external environment, provides a classic example. In homeostasis, stability is maintained as a result of self-regulation, operating through multiple layers of interacting systems; the maintenance of a fairly constant temperature requires continual dynamic adjustments. One billiard ball striking another thus is a problematically simple analogy for the causation of disease. It is more illuminating to understand the causation of disease in terms of the disruption of the mechanisms by which homeostasis is maintained; and recovery in terms of the re-establishment of the conditions for homeostasis (Billman 2020: 1). The same points apply *mutatis mutandis* where social systems such as the composition of a neighbourhood either maintain stability, or change in crucial respects over time. Stability or change within the social realm should be understood in terms of the interaction of mechanisms, rather than presupposed as a constant.

Tracing the connections between different regulatory mechanisms at different levels, and especially how their interactions will affect the distributions of variables, is extremely difficult. Often these mechanisms—even where they can all be identified—will interact in ways that are highly sensitive to small changes. Construction of models is crucial to help understand these interactions, but it is important to acknowledge that such models are by their nature simpler than the phenomena they are trying to model. It should be expected that any predictions derived from the model may diverge in important ways from what will actually happen. Particularly in the world of policy, for reasons I shall explain in due course, models are much more useful for scenario planning than for prediction. They can help researchers and policymakers to examine a range of ways that events could pan out, and plan or practise responses to these, even while it is acknowledged that the way things actually do pan out is unlikely to be exactly like any of these scenarios.

One important implication is that the world in which policymakers must act is in important respects unknowable. How citizens and other stakeholders respond to a policy is crucial for fixing what a policy means in practice, whether the policy incorporates a defensible balance of public values, and whether it will garner sufficient compliance to succeed. Policymaking must be understood as an improvisatory activity in which citizens are key partners. This understanding of the complexity of social processes needs to be allied with a conception of philosophy that focuses on doing rather than knowing, and is inspired by classical pragmatists such as John Dewey (1917) and William James, as well as contemporary pragmatists such as Kitcher (2012) in their call to re-situate philosophy within the problems of human life. Such a philosophy will be eclectic, synthesizing whichever perspectives and systems of disciplinary knowledge are helpful in informing action. We begin by introducing some broader concepts within complex systems theory, before examining their implications for thinking about the ethics of public health policy.

4.2 Parts, Wholes, and Complexity

Much progress in science since the Scientific Revolution has been aided by methodological reductionism—the assumption that the best way to understand things that seem complicated is by splitting them into smaller and smaller parts, studying each separately, and building back up to the understanding of larger entities. Methodological reductionism can be a successful strategy only for scientific problems with a certain structure—namely those where the parts that are identified and from which understanding is built up function in a way that is atomic, rather than being determined or modulated by their relations to other elements that make up the whole.

Many scientific problems can be tackled successfully through methodological reductionism. Much progress in science progress up to the middle of the nineteenth century came from focusing on what Weaver (1948) called *problems of simplicity*—scientific problems that allow everything other than one or two variables to be held constant. The later nineteenth century added the ability to tackle what he called problems of *disorganized complexity*—in which the interaction of a large number of variables that cannot be kept track of individually is treated statistically to allow a system as a whole to be seen to be behaving in accordance with statistical laws such as the second law of thermodynamics. However, by the time that Weaver was writing in 1948, it had become apparent that problems such as weather prediction, understanding how genes code for proteins, and why certain chemicals are carcinogenic, were resistant to being treated either as problems of simplicity or as problems of disorganized complexity. These problems of *organized complexity* required a different set of tools and techniques.

The following decades saw the rise of a variety of different disciplinary approaches to better understanding and theorizing organized complexity, including cybernetics, systems dynamics, and systems theory, which have now coalesced around the title of complexity science (Waldrop 1992; Pickering 2010; Rid 2016). Crucial to the study of organized complexity is the idea of a *system* as an assemblage of components that are sufficiently interrelated or interdependent as to be fruitfully conceived of as elements of a whole, and where the whole is greater than the sum of its parts. Not all sets of items will be described usefully as a system. Clearly, if the connections between the disparate elements are slight or adventitious, then there will be little to be gained by describing these as a system. In addition, there will be cases where even though the elements that are described as forming a system are indeed interrelated, the behaviour of the compound entity can be adequately understood 'bottom up', through understanding the parts individually, without needing to invoke any higher level or systemic principles.[1]

---

[1] Whether the whole is greater than the sum of its parts can be interpreted either as a metaphysical question, or as an epistemic question. When someone asserts that the whole is greater than the sum of

Complex systems are a subset of systems. The definition of what makes a system complex, and how to measure complexity, is itself no simple matter: there is a whole industry aiming to provide definitions of complexity, and agreement on how to measure it (Ladyman et al. 2013). For our purposes, the crucial thing is to think of complexity as a way of structuring assumptions about what the world is like, and thus what good inquiry, whether scientific or philosophical, should look like. As Arthur (2014: 3) puts it, 'Complexity is not a theory but a movement in the sciences that studies how the interacting elements in a system create overall patterns, and how these overall patterns in turn cause the interacting elements to change or adapt.' Herbert Simon provides an early, but still very helpful, intuitive characterization which provides orientation:

> Roughly, by a complex system I mean one made up of a large number of parts that interact in a nonsimple way. In such systems, the whole is more than the sum of the parts, not in an ultimate, metaphysical sense, but in the important pragmatic sense that, given the properties of the parts and the laws of their interaction, it is not a trivial matter to infer the properties of the whole. In the face of complexity, an in-principle reductionist may be at the same time a pragmatic holist.   (Simon 1962: 468)

Complex systems of this type are everywhere in the natural and social world, from climate and weather, to the spread and mutation of infectious diseases, to the workings of economies.

Complex systems are often nested one inside another. The concepts with which a scientifically literate person understands the world already contain within them the assumption of nested systems of parts and wholes, which go many layers deep. The human organism provides a vivid example. A human being is composed of around 100 trillion human cells, and is host for around 130 trillion more microbial cells. Cells of the human body are composed on average of billions of

---

its parts in a metaphysical sense, they assume that there is something extra, which has its own self-standing ontological heft, which comes into existence when the parts are connected in the right way. If the claim is made in an epistemic sense, there is no commitment to extra self-standing metaphysical entities being brought into existence: rather, it is claimed that the order perceived at a higher order is unpredictable or unexpected given the lower level structure. It is useful to use the higher level concept of the whole in explanations, because this allows certain things to come into view that would not be easily visible (or visible at all) by focusing only on the parts individually. Philosophers sometimes describe these two positions as strong and weak emergence, respectively (Bedau 1997).

Complexity theorists are nearly all weak emergentists. In their view, it is because of the arrangement and interconnection of the elements that features emerge at a higher level of organization, and it is in virtue of these higher order features that it makes sense to treat the assemblage of objects as a system. Doing so does not commit the systems theorist to the idea that there is anything additional or 'spooky' about the whole thus conceived, or that the whole could not ultimately be explained in terms of the behaviours of the elements out of which it is constructed.

molecules; each molecule is composed of multiple atoms; each atom is composed of multiple fundamental particles.

Each cell in the human body is a complex system, composed of a cell wall, cytoplasm, mitochondria, and nucleus; and with enough internal complexity to be able to reproduce themselves through mitosis. All this is possible because of the complex chains of molecules that are the building blocks of life. Cells can themselves be infiltrated by viruses, which hijack the cell's replication machinery to reproduce copies of the virus instead. Cells in turn make use of Dicer enzymes as part of antiviral defence. Ascend a level from cells, and we can think at the level of a tissue; for example, the epithelial tissue that lines the digestive tract, or the muscle tissue that forms the individual heart muscles.

Ascend another level, and we have whole organs such as the liver or gallbladder. Ascend another level, and we have organ systems such as the digestive system or the respiratory system. Ascend from the level of organ systems, and you reach the human organism. Ascend again and the individual can now be thought of as part of a family. Ascend again, and the family can be thought of as part of a community. Thus, although we may not often think of ourselves in these terms, an individual human being is composed of nested layers of systems, and is nested in further levels of larger systems.[2]

We readily see in cases such as communicable disease outbreaks how these different levels interact. The influenza virus is often spread by airborne transmission. An individual coughs or sneezes, and in doing so sheds virus particles that can then be inhaled or otherwise find their way into other susceptible hosts. Once inside the host, the virus enters susceptible cells and multiplies, colonizing the respiratory system, leading to wider systemic challenges for the human being whose respiratory system this is. This individual themselves becomes infectious, potentially infecting others. Families or whole communities may be laid low if the strain is virulent enough. And of course, each time the virus replicates there is a chance that it will mutate, and that this mutation will change its effects on different human systems—becoming either more or less virulent, and more or less able to out-compete other variants of the virus.

4.3 Stocks, Flows, and Models

Ideas of stocks and flows are helpful for understanding complex systems. The stock of a particular item is how much of it there is in the relevant system at a

---

[2] Moreover, the distinction between 'self' and 'other' is much murkier and more difficult to draw than common sense would usually suggest. This is not only because of the high percentage of microbial cells within the human body, but also because in some cases, such as gut microbiota, they are vital to normal functioning. For an accessible introduction to these questions, see Yong (2016).

particular time. The flow is the rate of change (positive or negative) in this stock. When you run a bath, the amount of water in the tub is the stock, while the flows can be positive (from the taps) or negative (if you pull out the plug) (Sterman 2001).

Thinking in terms of stocks and flows is useful in many areas of health policy. The current population of a country is its stock of people; a stock into which there are inflows from new births and immigration, and outflows from deaths and emigration. The number of patients waiting for treatment in an emergency department is also a stock, with inflows as more new patients arrive, and outflows as patients are triaged and sent on their way, or moved to another department of the hospital. The number of available doctors working in a country is a stock, with inflows as new doctors qualify or move to the country, and outflows as doctors retire, die, emigrate, or simply decide to work less.

Just as running or draining a bath takes time, so does increasing or decreasing stocks; and the rate depends on the net rate of flow. In some cases there will be significant delays between a decision to increase the rate of flow, and seeing any change in stocks as a result. For example, if a health system decides that it has a shortage of doctors, which it wants to meet by increasing numbers in medical training, then it will be several years from the initial decision to the increased number of practising doctors.

In complex systems, rates of flow (and thus stocks themselves) are subject to feedback loops: the size of an existing stock affects the size of that stock in the future. Feedback loops can either reinforce an existing rate of change, or can balance it. The thermostat on a heating system provides a simple example of a balancing loop: the system monitors the temperature in the room, causing the boiler to heat up if it is below the target temperature, and to shut off if it is above the target temperature. Infectious disease provides obvious examples of reinforcing feedback loops: one index case can infect several others, and each of these can infect several more, until there is a widespread outbreak. A similar dynamic is exploited when vaccination policies aim to ensure coverage of enough of the population to ensure herd immunity, i.e. that even if some individuals do become infected by the disease in question, it will be unlikely to spread because the average infected person will not encounter enough susceptible others during the period in which the disease is infectious for the disease to spread. Chapter 10 discusses communicable disease modelling in much more depth.

Feedback loops also provide a helpful way to think about the cultural influences on health and health policy. One among other factors that determine the rate at which young people will take up smoking is the rate of existing smoking amongst others they associate with, and the cultural expectations that accompany these factors. (Is smoking considered to be normal or even expected? Would one have to explain a decision *not* to smoke?) Similarly, as the number of doctors per capita falls within a poorly resourced public healthcare system, it may become increasingly difficult to keep doctors within the system—with doctors departing either to the private sector or emigrating to improve their life chances. As these examples and others that will occur to the reader indicate, in any real-world system that has a significant impact on population health, whether positive or negative, there will be multiple inflows and outflows to the system, and the rates of these inflows are affected both positively and negatively by factors that are internal to the system, and factors external to it.

The net result is that the relationship between flows and stocks within a large interconnected system is often difficult to map, let alone to predict or control over time. First, there may be a significant delay between the change in one stock, and that of another. Particularly when combined with patchy data, this may make it highly likely that any intervention will either be an under-reaction or an over-reaction. Second, where there are reinforcing feedback loops, and the increase in the stocks leads to further increase in the stocks (and potentially to increases in the rate of change), then what was initially an unnoticed problem can soon become an unmanageable one. As data on stocks and rates of changes will often be poor, these loops can easily escape notice. Third, it will often be the case that feedback mechanisms that serve to introduce balancing loops only work within certain limits: outside of those limits, the mechanism may change or reverse.

Making decisions to intervene in complex systems is to make decisions under uncertainty. When dealing with complex systems, it is illusory to think that decision-makers will be able to command a sufficient knowledge of the relevant facts such that they will never be surprised by what transpires. It is important to distinguish complex systems as they operate in reality from models of such systems. The earlier point about the sheer number of cells that compose a single human being was designed to show how unfeasible it will ever be to model a macroscopic complex system at a molecular level. For example, the most powerful existing weather models we have—such as the UK Met Office weather model—operate (at the time of writing) at a scale of 10km2 over the surface of the earth (Met Office 2020). Reducing these to a molecular scale would be utterly unfeasible.[3]

---

[3] As Borges reminds us, a map is useful only if it is more manageable than the territory of which it is a map: '... In that Empire, the Art of Cartography attained such Perfection that the map of a single Province occupied the entirety of a City, and the map of the Empire, the entirety of a Province. In time, those Unconscionable Maps no longer satisfied, and the Cartographers Guilds struck a Map of the Empire whose size was that of the Empire, and which coincided point for point with it. The following Generations, who were not so fond of the Study of Cartography as their Forebears had been, saw that that vast Map was Useless, and not without some Pitilessness was it, that they delivered it up to the Inclemencies of Sun and Winters. In the Deserts of the West, still today, there are Tattered Ruins of that Map, inhabited by Animals and Beggars; in all the Land there is no other Relic of the Disciplines of Geography' (Borges 1998: 325).

The yawning gap between complex systems and our models of them has important implications. For practical purposes, choices need to be made about (1) the factors within the system of interest that will be represented in models of it, (2) how to measure or approximate the values of these variables, and (3) the boundaries of the model. Each requires simplifications of a more complex reality, and thus introduces inaccuracies: things that are not included within the model of the system can and will affect factors inside the system we have mapped. Given the presence of feedback loops within systems, something that is either outside the model of the system, or within it but not accounted for by it, could come to have a very large effect within it. Those working in Chaos Theory have sometimes vividly illustrated this with the thought that the flapping of a butterfly's wings could, weeks later, cause a storm on the other side of the world (Lorenz 1963).

The upshot is that while there is no practical alternative to the use of models, even a sophisticated and well-tested model of a complex system should not be expected to predict accurately what is going to happen over the medium or long term. As Box put it, 'all models are wrong; the practical question is how wrong do they have to be to not be useful' (Box and Draper 1987: 74). Rather than thinking of models of systems as affording perfect control, it is better to think of their purpose as aiding scenario planning: helping to answer 'what if' questions, and allowing us to think creatively about what some of the implications of changes in one variable in the system might be.

4.4 A Complex Systems Approach to Public Health Policy

It should now be clear why it is a mistake to think that policy interventions are like bricks that can be made and tested to specification in one place, and then slotted into a structure elsewhere and expected to work as planned. The most important points to take forward from this brief introduction to complex systems are as follows. First, the extent of the interconnections of the systems that are relevant to government policies means that there is rarely a linear correlation between policy-level interventions, and improvements to individuals' lives. If, say, displaying healthy eating messages on one billboard has effect x, we should not expect displaying the message on ten or one hundred billboards to have an effect of $10x$, or $100x$.

Second, while some causal effects within these systems are obvious and direct, others are much more subtle, indirect, and difficult to detect. In particular, there can be delays in causation: shifting from government inspectors undertaking building fire safety inspections to a system of self-reporting might save money in the short term, but some years later could lead to problems with increased rates of fire. Similarly, banning smoking in public buildings may lead to wider patterns of change, e.g. reduction in smoking rates among smokers rather than simply getting the same amount of smoking to be undertaken outdoors. A particular trajectory may look obvious after the fact, but be far from but obvious at the time that decisions are made.

Third, it is vital to understand the system in which one is attempting to act. Points at which intervention will be maximally effective will often be counterintuitive; seemingly obvious changes may be ineffective or even counterproductive. Successful interventions need to take account of the broader systemic context: as Rutter et al. (2017: 1) put it, '[i]nstead of asking whether an intervention works to fix a problem, researchers should aim to identify if and how it contributes to reshaping a system in favourable ways'.

Thinking in complex systems terms alerts us to certain characteristic pathologies that systems can exhibit—what Meadows (2008) calls systems traps. Two of the most significant for our purposes are *policy resistance* and *success to the successful*. Policy resistance occurs where an attempt to intervene in a system through a 'quick fix' ends up not working because of a failure to anticipate its systemic effects. Thinking in systems terms requires a change of perspective in cases of policy resistance: it is no use blaming someone else, or unanticipated 'side effects':

> there are no side effects—just *effects*. Those we expected or that prove beneficial we call the main effects and claim credit. Those that undercut our policies and cause harm we claim to be side effects, hoping to excuse the failure of our intervention. 'Side effects' are not a feature of reality, but a sign that the boundaries of our mental models are too narrow, our time horizons too short.
>
> (Sterman 2006: 505)

Success to the successful operates where the system creates path dependencies that make it likely that those who are already doing well end up even better off, and those who are doing badly end up even worse off (Meadows 2008: ch. 5). The underlying mechanism and insight has been remarked on repeatedly in different policy arenas, but it is not always explicitly framed in terms of systems theory.

Tudor-Hart (1971) coined the term 'Inverse Care Law' to describe the fact that systems dynamics, if uncorrected, are likely to lead to the areas that have the greatest medical need having the least density of healthcare professionals. The explanation (as we will examine in more detail in Chapter 9) is that the social determinants of health imply that those who are already disadvantaged in other respects are more likely to suffer worse health, and that this worse health will in turn further exacerbate their other disadvantages. Thus healthcare need will tend to cluster in areas where there is greatest broader disadvantage. This will manifest itself in longer waiting lists and patients who are more difficult to treat; moreover, areas in which disadvantage is deeply woven are perceived to be less desirable to live in, and so healthcare professionals who have a choice are more likely to choose to practice in 'better' areas. The net effect is that, absent systematic attempts to equalize the resources available for care provision by need, we can expect an inverse care law to operate.

Observing an analogous systems dynamic, the sociologist Robert Merton coined the term the 'Matthew Effect' (drawing on the parable of the talents in Matthew 25:29), to explain how credit is often assigned in science: 'For to every one who has will more be given, and he will have abundance; but from him who has not, even what he has will be taken away.' Merton argued that systems of prestige in science tended to magnify the extent of contributions made by those who are already well known, while minimizing the extent of contributions of those who are relatively obscure; and that given that this process was iterative, led to a problematic path-dependence in which scientific prestige and resources were significantly determined by social position (Merton 1968).[4]

Brian Arthur (1989) noticed a very similar dynamic within the workings of markets, which in the economics literature is known as increasing returns. Economists had previously assumed that competition within a market creates diminishing returns—that after a certain point increasing market share becomes increasingly difficult, and so that competition within a market will lead to efficient allocations and low profits. Arthur argued that markets for intangible goods such as computer operating systems or social networking apps exhibit features such as network effects, high up-front costs but low marginal costs, and customers getting used to particular services and products. The combination of these features implies that there will tend to be increasing returns in such markets. Those who get ahead will tend to stay ahead, and there will also be significant path dependencies: the solution or service that comes to dominate may not be the best one.

4.5 The Normative Implications of Complex Systems

These features of complex systems will undoubtedly make a difference to the success of different strategies for intervention that policymakers could use, but what difference do they make to the ethical principles that should be used? Many philosophers and ethicists seem to think that the work of policy-relevant ethics can go on largely unchanged even once we fully acknowledge the extent to which public policy interventions are interventions into complex systems. I think that this is fundamentally mistaken. In order to understand why, we need to take a

---

[4] Merton's paper drew heavily on empirical research done by Harriet Zuckerman (1965), to whom he was later married. In a somewhat ironic confirmation of the Matthew Effect, her role has often been underplayed. In an attempt to rectify things, Merton added the following note to the 1973 reprinting: 'It is now [1973] belatedly evident to me that I drew upon the interview and other materials of the Zuckerman study to such an extent that, clearly, the paper should have appeared under joint authorship.'

brief detour to understand what is at stake, and why someone might wish to resist the idea that complexity has important implications for how to think about ethical principles in public health and public policy more broadly.

In framing the disagreement, it is helpful to begin from a distinction that Ackoff made between a problem and a mess:

> Managers are not confronted with problems that are independent of each other, but with dynamic situations that consist of complex systems of changing problems that interact with each other. I call such situations *messes*. Problems are abstractions extracted from messes by analysis; they are to messes as atoms are to tables and chairs. We experience messes, tables, and chairs; not problems and atoms... Managers do not solve problems; they manage messes.
>
> (Ackoff 1979: 99–100)

This distinction allows us to re-express the main claim of Chapter 3, in the light of a complex systems approach. A philosophical thought experiment makes a mess available for a certain kind of thought by not only stripping away a lot of the detail, but also by stripping away its dynamic and historical nature. It is a particularly stark example of reconstituting a mess as a problem. Ackoff's worry is that providing a lucid analysis of the problem described in the thought experiment does not necessarily tell you anything about how to manage a mess, both because the thought experiment is a simplification of the real-world problem, and because 'solving' the thought experiment does not give us insight into how the messes we face interrelate. As he puts it: 'The optimal solution of a model is *not* an optimal solution of a problem unless the model is a perfect representation of the problem, which it never is' (Ackoff 1979: 97).

One possible reason for disagreeing that complex systems theory has significant ethical implications for public policy would be to disagree about the facts. A philosopher might disagree about the facts about what causes what, and the degree of understanding, predictability, and control over these systems of causes that governments or agents more broadly have. In short, they might not believe that the social world is to a significant degree composed of complex systems. This is a possible position to espouse, but it is not one that gives credit to anyone who attempts to do so. Given the mass of empirical evidence from many years of complexity science, and the frequent and obvious failures of prediction and control within politics, the philosopher would need to overturn a large and diverse body of empirical research to make this claim plausible. It is the kind of claim that would make someone who made it look foolish rather than wise in their defence of philosophy.

A much more promising view for someone who wished to defend the status quo in philosophy would be to argue that *notwithstanding* the empirical reality of complex systems in human life and societies, philosophers should adopt an approach that is analytical and reductive, and isolates individual parts of systems and considers these parts in isolation from other parts—all the while acknowledging that in the real world any part that is isolated for the purposes of analysis affects and is affected by other elements of the system.

There could be various reasons for adopting an atomistic mode of analysis to what is admitted to a part of an interconnected system that constitute a whole. First, it could be argued that once we better understand the individual parts, then we can put everything back together again; it is just too hard to think about all of the system at the same time. Second, and more radically, someone might wish to claim that while our lived world is indeed composed of messes rather than problems, the task of philosophy is nonetheless to study problems rather than messes, because at a deeper ethical level the world is one of principles and problems rather than messes. I will consider each in turn.

4.5.1   The Usefulness of Abstraction

The claim that it is often useful to isolate one part of a complex system to better understand it is very plausible—and, as we have seen, modelling of complex systems itself nearly always involves simplifications and decisions about where to draw model boundaries. Thus, those working in complexity science will agree that it is sometimes or often essential to understand the parts in order to better understand the whole. But such abstraction needs to be done for the right reasons, and in the context of the broader system one is aiming to understand. If very little attempt is in fact made to integrate the insights gained from looking at the parts separately into an understanding of the whole, then the appeal to the need for atomistic abstraction may be more plausibly thought to be a rationalization rather than a good faith attempt to understand the whole.[5]

The way that the language of parts and wholes is typically used may prime us to misunderstand the nature of complex systems. As Arthur Koestler put it:

> wholes and parts in this absolute sense do not exist anywhere, either in the domain of living organisms or of social organizations. What we find are intermediary structures on a series of levels in ascending order of complexity, each of which has two faces looking in opposite directions: the face turned towards the

---

[5] One of the under-remarked features of the back-and-forth of academic debate is the way in which as the literature is defined and refined, the purpose of initial abstractions can easily become lost. What was proposed as a means becomes an end. A process of forgetting sets in, and the simplified abstraction becomes reified: academic careers can be made by working within the abstract model, and there is often little requirement or even incentive to think about the larger question of the whole. Dominant methodological approaches within the discipline may encourage this reification: as Chapter 3 explored, the types of methods developed by philosophers since Nozick are dependent on constructing simple and acontextual scenarios and trading intuitions about these.

lower level is that of an autonomous whole, the one turned towards the higher levels is that of a dependent part.   (Koestler 1970: 135–6)

Koestler attempted, with limited success, a linguistic reform, introducing the idea of the *holon* as a stable sub-whole. On this view, each holon was both a constituent of other holons, and is itself constituted by holons: cells, tissues, organs, organ systems, human beings, families, and communities would all be holons. While it will often be useful to focus on just one or two levels for the purposes of a particular piece of analysis, it is important that each can be both part and whole depending on our interest and level of analysis.

Complex systems theorists argue that deep facts about biological and social reality explain why biological and social systems are better understood through the idea of nested complex systems, rather than neatly separable parts and wholes. To fail to take these interconnections seriously is to fail to understand the entities abut which one is attempting to theorize.

Here is a sketch of some of this story. The processes of evolution—random variation, genetic inheritance, and competition—entail that organisms tend to be well adapted to the ecological niches in which they find themselves. Those that are not do not survive to pass on their genes. These processes of competition occur simultaneously at a variety of scales from the micro to the macro: it is not just a matter of, for example lions evolving to be better hunters of gazelles, and gazelles in their turn evolving to be better at avoiding being eaten by lions, but of whole ecosystems. Many ecological niches (such as gut microbiota) involve being inside another organism. Mutualism is common throughout nature, which can then lead to further evolutionary advantages to deepening mutualism. In short, the struggle for survival creates conditions that favour interconnection and interdependence.[6]

Biological systems are argued to have a nested hierarchical structure, and these systems are partially decomposable (that is to say, we can focus on one level for analytical purposes, and leave the others on one side), because hierarchically structured organisms can both evolve much more quickly, and are more resilient against failures of DNA replication and failure of parts.[7] A similar style of explan-

---

[6] For an introduction to niche construction theory, see Laland et al. (2016).

[7] Simon (1962: 470) explains this via a famous parable of two watchmakers, Hora and Tempus, who each make excellent watches: 'The watches the men made consisted of about 1,000 parts each. Tempus had so constructed his that if he had one partly assembled and had to put it down—to answer the phone say—it immediately fell to pieces and had to be reassembled from the elements. The better the customers liked his watches, the more they phoned him, the more difficult it became for him to find enough uninterrupted time to finish a watch. The watches that Hora made were no less complex than those of Tempus. But he had designed them so that he could put together subassemblies of about ten elements each. Ten of these subassemblies, again, could be put together into a larger subassembly; and a system of ten of the latter subassemblies constituted the whole watch. Hence, when Hora had to put down a partly assembled watch in order to answer the phone, he lost only a small part of his work, and he assembled his watches in only a fraction of the man-hours it took Tempus.'

ation also allows systems theorists to explain why social reality is also composed of nested systems. Human babies are unable to develop into toddlers, children, and then adults without extensive care over a long period of time from those playing a parental role. A family structure of some kind thus seems to be a necessary feature of human life. The child's ideas of the self, and of identity, and of morality are formed in dialogue with these carers; and so the child from the beginning comes to view themselves in relationship to others. Those who nurture children, and socialize them in these ways are themselves drawing on a background of culture and language that is shared more broadly; the learning that is accreted in language and culture is so valuable and short-circuits so many false ends, that it would be massively expensive to do without this. We are both biologically and socially embedded in complex systems.[8]

4.5.2   Is Moral Reality Simple?

Even the claim that it is *useful* for philosophers to treat parts in isolation from the wholes that they constitute faces significant challenge, given that the decomposability that is seen in biological and social realms is always partial. Moreover, it makes the appropriateness of abstraction a more contingent and empirical question than many philosophers will be comfortable with. So some philosophers might be tempted to fall back to a much less empirically accountable claim—namely that the facts about complex systems, and the corresponding bodies of biological, psychological, and social theory, have no direct ethical implications.

Each time their phone rings, Horus will have to reassemble an average of five parts; whereas Tempus will have to reassemble many more (the average number will depend on the chances of interruption; but it could be up to 999). The net result is that if the chance of being interrupted when assembling any individual piece is 1%, then it will take Tempus 4,000 times longer to make each watch than Hora. One implication of the parable is that complex creatures that have a hierarchical structure with nearly decomposable elements will evolve much more quickly than ones that are not, and will be more resilient when they do evolve. Indeed, Simon argues that there would simply not be enough time for complex creatures like us to have evolved unless a hierarchical structure had been selected for.

[8] Habermas provides a helpful anthropological account of the function of morality, which links the requirement for norms requiring interpersonal consideration to the possibility of us becoming the kinds of being we are: 'I conceive of moral behavior as a constructive response to dependencies rooted in the incompleteness of our organic makeup and in the persistent frailty (most felt in the phases of childhood, illness and old age) of our bodily existence. Normative regulation of interpersonal relations may be seen as a porous shell protecting a vulnerable body, and the person incorporated in this body, from the contingencies they are exposed to. Moral rules are fragile constructions protecting *both* the physis from bodily injuries and the person from inner or symbolical injuries. Subjectivity, being what makes the human body a soul-possessing receptacle of the spirit, is itself constituted through intersubjective relations to others. The individual self will only emerge through the course of social externalization, and can only be *stabilized* within the network of undamaged relations of mutual recognition' (Habermas 2003: 33–4). For more on attempts to provide a naturalistic 'bottom-up' explanation of how ethical norms and modes of reasoning could have developed without reference to a moral reality external to human beings, see Kitcher (2011b) and Tomasello (2016).

As an example of this kind of strategy, Cohen (2009) argues that justice itself is ideal, and that it is a mistake to think that it is a desideratum for a theory of justice that it should be helpful in guiding action in real-world contexts. Cohen argues that we should distinguish theories of justice from theories of optimal regulation. Theories of justice proper need to rest on fact-free principles; and thus must be true in all possible worlds. Theories of optimal regulation apply theories of justice given empirical constraints such as human nature and the level of technological development. According to Cohen, political philosophy should be concerned only with justice; principles of optimal regulation fall outside of the domain of philosophy.

Analogously, it would be fairly easy to acknowledge that the world as we experience it is filled with complex systems, and to admit that we need to take this complex and systemic nature into account when ethical principles are being applied, but not when we are investigating their deep structure. I shall describe this view as a High Theory approach (Arras 2016). High Theory would be a way of keeping ethics safe from complex systems, but at a significant cost. The philosopher would remain master of a terrain that is not easily criticized (or even discussed) from outside of a fairly narrow set of philosophical assumptions. This view can no doubt be put in a way that is coherent; that is to say, that it need not involve self-contradiction. But it nonetheless seems to me deeply mistaken.[9]

I acknowledge that this is not a debate that will be resolved by a single argument; what we see is two different research paradigms. I have two main concerns about High Theory views of ethics, which the rest of this chapter will develop, and the rest of the book will exemplify. The first concern is that High Theory approaches are not useful in practice. I start from the view that it is the *real world* in all its complexity that policymakers are attempting to change. Acting in the real world requires us to abstract and model in various ways so that we can better deliberate: abstraction thus serves the purpose of *reculer pour mieux sauter*. The High Theorist, on the other hand, tends to think that the real world provides but seams of philosophical ore, which need to be mined and refined before they can be made properly philosophically useful.

High Theory approaches embody a conception of philosophy that emphasizes the construction of and resolution of *puzzles*, rather than problems or messes. This risks turning philosophy into a narrow elitist pursuit that is of little interest to anyone outside of a narrow circle of academic philosophers (Kitcher 2011c). One reason that the challenge of complexity may have been overlooked in philosophical work that bears on public policy is that there are relatively few philosophers who have significant experience of public policy. When those who

---

[9] I am with Dewey here: 'Philosophy recovers itself when it ceases to be a device for dealing with the problems of philosophers and becomes a method, cultivated by philosophers, for dealing with the problems of men' (Dewey 1917: 65).

develop the theory that is supposed somehow to inform practice never have practised and never aim to put their theories into practice, and are aiming to impress those who are similarly situated, then it is easy for the theory to become unhelpful. As has been studied within the broader literature on research organization, where the incentives on academics are largely to produce work that will impress other academics within their discipline, a significant gulf can open up between research that is deemed to be excellent within the discipline, and research that allows progress on problems that are visibly important from a broader trans-disciplinary or social perspective (Stilgoe 2014).

The second, and deeper, problem is that the High Theory approach misunderstands the nature of our ethical concepts and practice—where they come from, and how they relate to the world. As the next section explains, ethics needs to take account of the fact that the concepts through which we interpret the social world, and the ethically relevant facts that are created by these, are partially socially constructed. The project for the ethics of public policy is not best thought of through how to apply unchanging ethical values to a changing and uncertain set of complex systems, but how to understand the complex systems that constitute the realm of social policy, in order to do better in designing ethically defensible policies. Our ends and the means of achieving these are not something external to our developing understanding of the problem, but are part of the inquiry. Because of this there is a core element of creativity and fluidity to ethical thinking.

4.6 Performativity in Complex Systems

It is often taken for granted that what one is attempting to explain or predict is unaffected by our attempts to do so: predicting a solar eclipse does not affect its timing. However, in systems in which the behaviour of human beings plays a part, 'public definitions of a situation (prophecies or predictions) become an integral part of the situation and thus affect subsequent developments' (Merton 1948: 195). Merton underlines this point via a parable about a bank which is in good financial health, but becomes insolvent as a result of a rumour. The rumour causes more and more of its customers to want to withdraw their money; queues get steadily longer, and the bank does not have enough cash at hand to fulfil all these requests. There is a run on the bank, and it collapses. The bank *becomes* insolvent, in a self-fulfilling prophecy: a belief about a situation comes to be true because people act as if it is (Biggs 2011).

Complex systems in which human beings play a part exhibit not only self-fulfilling prophecies, but also a wider range of ways in which agents incorporate expectations of each others' behaviour into their own actions. How the human beings who partially compose a system interpret elements of the system, and how easy or difficult the expectations of others make it for them to get what they want, will alter the behaviour of that system. I describe this broader feature of complex systems as their *performativity*. Performativity is a well-known phenomenon in economics, leading to a whole literature on the extent to which economic models and predictions influence markets—pithily summed up by the economist Charles Goodhart, who was then an advisor to the Bank of England, in what has come to be known as Goodhart's law: 'Any observed statistical regularity will tend to collapse once pressure is placed upon it for control purposes.'[10]

Performativity can work both for good and ill. If you assume that someone is hostile to you, and then act on that basis, and they then respond to your apparent distrust, and you then take their response as evidence of their hostility, then things can easily escalate into a full-blown enmity in which it is *true* that the other person is hostile to you, even though this is true only because of your initial false assumption. Start with the assumption that the other is trustworthy and likeable and act on this basis, then the cascade of behaviour may go the other way; the other returns the openness and compliments, which are interpreted as signs of friendliness and as the cause for further friendly moves.[11] Ultimately, we are not spectators or detached scientific investigators of social reality, but the actors who are creating the show.

William James gives a seminal analysis of the ethical implications of performativity in his 1897 essay, 'The Will to Believe'. James argues that while we usually start from the thought that we should only believe something if it is true, there is a vast and vital class of cases where whether something is (or will be) true depends on whether we (and others) are willing to act on the basis that it is. We see this in cases such as friendship, but also in the construction of social reality more generally (James 1897).[12]

---

[10] The changing role of volatility (where volatility measures the extent of price changes within a market) provides a vivid example. Back in the 1950s, some traders and economists thought it would be useful to start measuring the volatility of financial markets, and such measures gradually gained acceptance. By 1993, the Chicago Board Options Exchange's Volatility Index (VIX) was launched, which allowed traders to make bets on future volatility. By 2018, betting on volatility had become so popular that it had started to destabilize markets around the world, leading to massive losses. Measuring, and betting, on volatility had greatly increased volatility. See Wigglesworth (2018) for more on this.

[11] Easwaran relates the following parable: 'We have a story in India about two men, one high-minded and generous, the other very selfish, who were sent to foreign lands and asked to tell what kind of people they found there. The first reported that he found people basically good at heart, not very different from those at home. The second man felt envious hearing this, for in the place he visited everyone was selfish, scheming and cruel. Both, of course, were describing the same land' (Easwaran 1986: 66).

[12] The idea of social reality is contested, but for our purposes we can take it to be shared social expectations about behaviours, meanings, and norms that are relied on in planning and action. Social reality overlaps partially, but not completely, with reality as it is investigated in natural sciences: for example, the progress of the seasons is both an important aspect of social reality and is also readily explicable in terms of the way that the axis of the Earth is tilted and the effects that this has on how much sunlight falls on the northern and southern hemispheres at different times of the year.

Few things are as important for day-to-day life as money, but yet money has become ever more obviously and purely performative in its social meaning and value. The earliest widely used currencies were commodity currencies, which were taken to be valuable because of the materials (and the quantity) of materials from which they were composed—as when the worth of a gold coins derives from the fact that it is a certain weight of gold. Later, representative currencies were introduced: coins and banknotes were taken to represent, and be exchangeable on demand for, a certain value of commodities.

The twentieth century saw the rise of fiat currencies, where the value of the money comes neither from the materials from which it is made, nor from the fact that there is a legal right to exchange it for commodities, but has value as a medium of exchange because a government stands behind it. The shift to fiat (and the twenty-first-century rise of cryptocurrencies) has made clear that money need not have any physical form at all. What makes something money at all, and what other goods it can be exchanged for, depends on the willingness of others to treat it as money, and alters according to the dynamics of social systems. This much becomes painfully obvious in circumstances such as the hyperinflation in Germany in 1992–3, or Zimbabwe in 2008–9; but even in more ordinary circumstances the exchange rates of fiat currencies are continually altering on the basis of bets on the confidence of different economies.[13]

The classification and diagnosis of diseases also shows significant elements of social construction. Disease classification (nosology) clusters diseases into categories that are at least partly a matter of convenience: as we discover and know more, they are reconfigured in response to the demands of utility. The end result is that disease classifications are path dependent, and often not a matter of cutting nature at its joints, but of drawing lines and categorizing for pragmatic purposes (Aronowitz 2001).[14] The way that diseases are described and classified affects not only what healthcare professionals look for in making a diagnosis, but also the ways in which patients need to present themselves if they are to receive a diagnosis and further treatment.

This has particularly important implications such as Chronic Fatigue Syndrome (CFS), for which there is no gold standard test, and doctors typically make a diagnosis by elimination only after a variety of other possibilities have been ruled

---

[13] For a history of money, see Weatherford (1997).

[14] This is most obvious in the case of psychiatric illnesses (see, for example, Zachar and Kendler 2017), but is also the case for diseases such as dementia. Dementia patients who present with certain disease features are fitted into existing categories that clinicians use, while clinicians are also aware that within each category there is likely to be heterogeneity. A patient may initially seem to fit a particular category, but over time, as a clinician sees more and more similar patients, they may come to judge that instead of being a phenotypic variation within that category, it might actually be a different disease (Morris 2000). Or further research might suggest reasons to consider two previously disparate diseases to be part of a common continuum (as Jellinger and Korczyn 2018 explore for dementia with Lewy bodies, and Parkinson's disease dementia).

out. Dumit (2006) examined how CFS sufferers in the USA used internet discussions as a way of working out how to present themselves in a way that would allow a diagnosis to be made that would assign them to a recognized disease category (either by presenting with depression, or via actually gaining recognition for CFS), and learning how to appear ill in the 'right' way that met the expectations of the system. One patient revealed that 'When my disability came up for review, my doctor told me to not dress nicely, not shower, and most of all, not bring my computer printout of all my symptoms, treatments, tests and doctors that I was maintaining then' (Dumit 2006: 585). The presented symptoms of CFS thus depend partly on what is expected within a health system to gain a diagnosis.

These are just few examples of the myriad ways in which social construction and performativity make a difference to our lives. The fundamental challenge for High Theory positions that aim to treat moral truths as something wholly separate from us, and which need to be somehow applied in order to decide what to do, is thus that the social reality to which we would be attempting to 'apply' these truths is not fixed separate from what we do, but is partly constituted by it. Vaccination policy, public trust in use of healthcare data, and social equality each show this dynamic clearly.

High rates of childhood vaccination are vital in order to prevent outbreaks of measles and other contagious diseases. In the case of measles, a vaccination rate upwards of 90 per cent (and preferably 95 per cent) is required to prevent outbreaks. Where the vaccination rate begins to fall significantly below this, governments have a choice about whether to make vaccination mandatory, and if so what penalties to apply to vaccine hesitancy or refusal. Whether making vaccination mandatory is a good solution to this problem, and whether it would even increase vaccination rates, depends to a significant degree on how citizens interpret the policy. Mandating vaccination might increase both vaccination rates and public confidence in vaccination, if citizens are moved by the dangers that would come from the loss of herd immunity, and feel that it is reasonable to require everyone to play their part in protecting the common good. Conversely, making mandatory a vaccination policy that was previously supported on a voluntary basis may act to reduce public confidence in it, and if the coercion is perceived to be heavy-handed or unmerited, may increase vaccine refusals.[15]

What counts as a violation of privacy is partly constituted by social norms and expectations. Obligations of privacy and confidentiality are ethically interesting in that while they are clearly of the first importance within healthcare, it is also the case that they are highly malleable (norms about privacy and confidentiality can differ greatly between contexts, and between states), and that what individuals choose or consent to makes a significant (but not total) difference to the

---

[15] Chapter 10.4 looks at the dynamics of vaccination in more detail.

obligations that others face. Reasonable expectations create a set of normative defaults, which then need to be overridden if the default is not going to happen. Such expectations play a *constitutive* rather than an *instrumental* role in creating obligations: it is not that clinicians merely should have regard to the fact that patients are likely to be surprised or angry if their data is used in certain ways, but also that these expectations play an important role in making it the case that there are certain obligations in the first place (Nissenbaum 2009; Rumbold and Wilson 2019; Taylor and Wilson 2019).

Being treated by others as an equal is a vital pre-requisite for self-respect and self-esteem. Yet the conditions under which it is reasonable to think of oneself as an equal to others, or to feel slighted by not being treated as an equal, are largely or completely socially constructed. Constructs such as 'race', social class, and caste can make profound differences to the life chances of those who find themselves labelled as part of a less-favoured group, despite the fact that these delineations amount to nothing over and above a collective willingness to judge and to enforce judgements on this basis. Power is crucial to these dynamics. Filling out what it is to treat individuals as equals within a modern healthcare system thus requires engaging in the kinds of detailed contextual understanding of the kinds of ways that equality can be undermined and what to do about it (for example, properly understanding mental health stigma, and how to counteract it), which must start from and work within the social construction of reality.[16]

4.7 Conclusion

The idea that we can improve ethical thinking by focusing on much simpler models of the real world, and then directly transferring insights back from the simpler model to the real world, has a long pedigree. Its cogency as a strategy depends on certain assumptions both about what the real world is like, and also about the solubility of the challenge of external validity. The last two chapters have argued that looking more closely at the world in which policymakers operate shows that these challenges are insuperable, and a complex systems approach needs to be adopted. Chapter 3 made the case that even if we assume that ethical and social reality is fixed separately from us, the problem of external validity would be a deep threat to the plausibility of top-down approaches.

This chapter has pushed a more radical critique: the ethical and social reality that policymakers need to navigate is composed of agents who themselves are trying to anticipate and respond to each other's (and policymakers') actions. This makes the decision problems they face at least partially dependent on people's

---

[16] Chapter 9 discusses these questions in more depth.

attitudes and judgements, and not even in principle decidable in the abstract for all cases.

As Part II (and in particular Chapter 7) will explore, the solution is not to give up on ethical theory, but to rethink it as a way of exploring the space of advantages and disadvantages of different ideals and ways of making trade-offs. Theorizing well in the abstract can give us a body of ethical thinking that can be drawn on creatively—even if not 'applied' in any obvious way—in addressing concrete problems.

PART II

BEYOND THE
NEGLECTFUL STATE

An Ethical Framework For Public Health

Introduction to Part II

The analysis undertaken in Part I showed that a particular model for making public policy interventions and of doing ethics, which focuses on simplification and
abstraction, is of little use for guiding policy. The idea that it is fruitful to devise
blueprints for what should be done, which then be 'applied' fundamentally mis-
conceives the nature of the task, given the challenges that complexity and, in particular, performativity, pose.

This analysis focused on ways that public policy should be done, rather than
the goals that should be pursued, or on what would count as a good society. So
Part I—although it will be challenging for some with particular views of how
public policy or philosophy should be done—need not have divided readers along
left-wing vs right-wing, or what is referred to in the USA as liberal vs conserva-
tive, lines. One indicator of the ecumenical nature of complexity approaches is
the sheer range of their precursors and fellow travellers—including market enthu-
siasts such as Hayek and the Austrian School of economics (Hayek 1967; Rosser
2015); game theorists such as von Neumann (1966) and Schelling (2006); work
on institutional diversity and common property regimes (Ostrom 1990, 2010;
Lewis 2017); the ecological economics of the landmark *Limits to Growth* report
(Meadows et al. 1972); and Jane Jacobs's work on cities and town planning (Jacobs
1961: ch. 22).

Each of these figures would be sympathetic to the idea that it is very easy for
top-down initiatives that fail to take into account systemic connections and feedback loops to make things worse. However, acknowledging that it is easy to intervene badly does not tell us anything about what a good society will look like, nor
how to design and implement interventions that work well. It will be no surprise
that the answers to these questions about the good society will split along political lines.

On the political right—and particularly those who are influenced by Hayek (1944)—the unknowability and unpredictably that comes from complexity and performativity is often thought to imply that it is difficult, if not impossible, for government interventions to succeed in improving the lives of citizens, and that it is better to leave things to be determined by the market.

I shall make a few brief remarks here in response to this worldview—though the longer reply will take up most of the rest of the book. While there are many instances in which public policy fails because of failure to attend to systemic inter-actions, it is simply mistaken to draw the conclusion that complexity approaches provide blanket support for government inaction. One obvious reason is that there is a range cases in which the mechanisms that lead to disease and ill health, or conversely to health and well-being, are well understood, and there are many interventions that have external validity and can be scaled cost-effectively to the level of a society. Examples would include childhood vaccination schemes, requir-ing drivers and passengers to wear seatbelts, and banning tobacco advertising.

A broader point is that the theoretical reasons that economists advance for thinking that, under certain assumptions such as perfect information and zero transaction costs, the deliverances of a market will be Pareto optimal hold only under conditions of equilibrium. The starting-point for complexity approaches is that economies are in fact rarely in equilibrium. As Arthur puts it in setting out the stall for complexity economics:

> Agents are not all knowing and perfectly rational; they must make sense of the situations they are in and explore strategies as they do this. The economy is not given, not a simple container of its technologies; it forms from them and changes in structure as this happens. In this way the economy is organic, one layer forms on top of the previous ones; it is ever changing, it shows perpetual novelty; and structures within it appear, persist for a while, and melt back into it again.
>
> (Arthur 2014: 2)

Looked at from the perspective of a rigorous complexity approach, competition in an economy is no more likely to lead to results that are optimal from an eth-ical perspective than is the process of natural selection (Sloan Wilson and Gowdy 2015; Sloan Wilson 2016). Without firm government intervention, the inter-related systems that compose society will often act to exacerbate existing inequalities and increase the risks to those who are most vulnerable. In such cir-cumstances, government inaction is not neutral, but amounts to allowing the vulnerable to come to harm.

The lesson that all policymakers should draw, regardless of their political per-suasion, is that they need to be clear about (1) the states of affairs they are aiming to promote, and the values that they are guided by, (2) how to reconcile conflicts between these values that arise both in principle and in practice, and (3) how the particular contexts and systems within which their public policies need to be enacted make some interventions more promising than others as a way of reconciling these values.

Both those who are in favour of interventionist public health policies, and those who wish to resist them, need to defend a set of values, explain how to reconcile these different values, and to explain how and why these values are better served by some kinds of policy than others. Such accounts need to combine normative claims about what a good society would look like, normative constraints on the kinds of policies that it is legitimate for governments to enact in order to get there, with a sophisticated empirical understanding of the ways in which different interventions would *in fact* impact on these values, given systemic interrelations and performativity.

When it comes to public health policy, at the deepest level, this debate is about the value of health, and how this value should be reconciled and traded-off against other potentially competing values such as individual liberty, autonomy, and equality. The central argument of Part II of this book is that the right balance to strike is one that places a very important weight on protection and promotion of public health, but does so in a way that is grounded in respect for individual rights. Part II focuses on clarifying the values that should guide public health, and how trade-offs should be made in principle. This involves making a number of general claims, which clarify the kinds of arguments that are often advanced either in favour of or against public health, and making new arguments.

The scope and nature of these arguments should be understood in the light of the arguments about the role and nature of philosophical argument in Chapter 4. Philosophical argument in ethics as elsewhere requires simplified representations, which convert what would otherwise be an intractable mess into something that can be considered and reflected on in thought. It is important not to confuse the model with the messier and more complex reality of which it is a model.

A thought experiment is a model; but so are analogies and stories. All methodologies for ethical thinking face the kinds of challenges of internal and external validity that Chapter 3 laid out in the case of thought experiments, so it would be futile to be against the use of models in ethical thinking. The takeaway from Part I is not a call to abandon thought experiments and other forms of abstraction, but a call to use them responsibly. Using them responsibly means accepting that the kinds of methods available in ethical philosophy are—at best—fallible ways of constructing simple models that map rather imperfectly onto the world as we experience it. It also requires accepting that the 'ethical reality' that ethics aims to provide insights into is in part a human construct, and so even where ethical inquiry does lead to consensus or to new insights, this may be best thought of as a kind of invention rather than a discovery.

The analysis in Part II aims to render perspicuous the ways in which certain arguments and ethical considerations do, or do not, bear on the activity of public health. Chapter 5 analyses the extent to which it is problematic for the state to improve health by interfering in individuals' lives. While public health policies are often accused of being paternalistic, or to show the 'Nanny State' in action, things are actually rather more complex, and complaints about paternalism in public health policy are much less convincing than is often thought. First, it is deeply problematic to pick out which policies should count as paternalistic; at best we can talk about paternalistic justifications for policies. Second, two of the elements that make paternalism problematic at an individual level—interference with liberty and lack of individual consent—are endemic to public policy contexts in general and so cannot be used to support the claim that paternalism in particular is wrong. It concludes that instead of debating whether a given policy is paternalistic, it would be better to ask whether the infringements of liberty it contains are justifiable, without placing any weight on whether or not those infringements of liberty are paternalistic. In doing so, it becomes apparent that a wide range of interventionist public health policies are no more problematic than a range of other government measures that impact on liberty and usually go uncontested, such as collection of taxes, regulation of advertising to ensure accuracy, and laws to prevent excessive noise at night.

Chapter 6 goes beyond defusing bad arguments against the legitimacy of interventionist public health, to make a positive case for it, by reframing the project of public health within a rights framework. It argues that there is a right to health, and this entails that individuals have a right to public health. Given that there is a right to public health, the state should undertake to reduce health risks. If it does not take easy steps to reduce risks to health, and as a result allows significant numbers to come to harm or even death, then it violates individuals' right to public health, and should be criticized as a Neglectful State. The ethical challenge of public health policy is therefore not the one-sided one of avoiding Nannying, but the more complex task of steering a course between Nannying and Neglect. Avoiding Neglect may involve restricting liberty in various ways. Though it might seem paradoxical to argue that an individual right to public health could justify restricting liberty, the chapter argues that this is an accepted implication of other rights such as the right to security, and so that there is in fact no paradox.

Chapter 7 addresses how to prioritize public health policies within the context of the right to health approach. The touchstone of this approach is that public health interventions need to be justifiable to individuals. For example, it is not enough merely to cite the size of the population health gain, if the intervention is one that involves an interference that would seem disproportional to most of those it affects. Designing approaches to prioritization that are adequately justifiable to individuals can be extremely difficult. One tool for clarifying the problem, which has been widely explored in the philosophical literature, is the idea of a *claim*—where the strength of an individual's claim depends on features such as how badly off they are, their capacity to benefit, the time at which their need arises, and whether the bad that will befall them is certain or merely possible. The chapter argues that it is mistaken to think that there is a single and uniquely correct way of measuring claims. This means that approaches to prioritization need to be pluralistic, and need to reflect on the measures most appropriate for a particular policy challenge.

The framework set out in these chapters is explored more fully in Part III, which provides a detailed analysis of three central arenas in which policymakers will need to combine an appreciation of competing values with an understanding of the nuances of systemic causation—respectively, responsibility, health inequalities, and communicable disease.

5

Paternalism, Autonomy, and the Common Good

Infringing Liberty for the Sake of Health

5.1 Introduction

Government interventions such as speed limits on roads, childhood vaccination programmes, regulation of toxic chemicals, and bans on smoking in public places improve population health. Nonetheless, concerted public health action by governments is often treated with suspicion. Public health sceptics challenge both the assumption that promoting population health is an effective way of promoting the common good, and the moral legitimacy of government interventions that interfere with individual liberties in the course of promoting the common good.

When such concerns rise to a philosophical level, they are most usually articulated through an interpretation of the proper role of governments, which places respect for individual autonomy at its core. This approach—call it the 'autonomy first' view—takes it to be axiomatic that it is difficult to justify state interference in the lives of competent adults unless the behaviours interfered with are compromised in terms of their autonomy, or would wrongfully infringe the autonomy of others. State interference in other circumstances is argued to be either counterproductive or wrongful, or both.[1]

Autonomy first approaches are united in thinking that public health activity is legitimate only where something that has gone wrong, autonomywise, with the behaviours or choices interfered with. The focus of justification for state public health activity must either be on showing that certain choices are not adequately autonomous and are thus fair game for interference (for example, those made by addicts about their drug of choice), or that the behaviours involve wrongful infringement of others' sovereign domain (for example, smoking in an office environment). Either way, it is assumed that individuals have an entitlement that their adequately informed and adequately voluntary decisions not be interfered

---

[1] Within bioethics, such a view is widely attributed to Mill on the basis of a few passages from *On Liberty*. However, Mill's actual position is rather more complex (see Jennings 2009; Claeys 2013).

with unless the interference is necessary to prevent violation of rights or rectify existing rights violations (I shall call this the *non-interference principle*).

Some recent influential positions, such as Thaler and Sunstein (2008), aim to avoid the alleged ethical problems with state interference by deploying only 'nudges' in public health policy, where nudges alter the framing of situations, but not the substantive costs and benefits attached to choices—as when a school canteen chooses to place the fruit in a more salient position than the less healthy puddings. However, the attempt to pursue public health policy without interfering with liberty or autonomy faces a fundamental challenge. Much if not most socially controllable health harm is caused in circumstances in which there is neither an obvious failure of choices to be autonomous, nor an obvious wrongful infringement of one person's rights by another. As Chapter 1 explored, the rise of obesity has a large number of contributory causes, many of which are distal rather than proximal. For example, a rise of processed foods; a decline in cooking skills; a shift towards designing spaces around car ownership and a corresponding decline in active transport; and a shift to more sedentary forms of work, as well as potential reinforcement effects from the rise of obesity itself (Butland et al. 2007).

Obesity is harmful in the sense of setting back people's interests in being healthy. But this harm has causes that are structural, diffuse, and multilayered. There are few if any individuals or institutions that act with the aim of increasing obesity, and the contribution of any given individual or corporation to obesity rates will usually be small or negligible. So while the harms are significant, there need be no sense of wrongful agency and no agent who can be singled out as 'the' cause of obesity. Rather, we are more likely to find that at each node of the network, agents (whether individual or corporate) act in ways that make sense given the constraints upon them. If no one can be shown to have acted wrongly, it is initially difficult to see how there could have been a wrongful infringement of autonomy. It is also far from clear if individual food choices are sufficiently non-autonomous to allow state action either.

This raises a fundamental question about the usefulness of the 'autonomy first' approach as a basis for public health policy once it is noted that complex and multilayered systems of causation are the rule rather than the exception in public health.[2] Much of the burden of disease results from structural processes that are contributed to by many people, and which are not intended to cause harm.

---

[2] Such problems are also common outside of the realm of public health. Climate change and global supply chains provide good examples. Sinnott-Armstrong (2005) examines how difficult it is to make the case that an individual action such as going for an unnecessary Sunday drive merely for pleasure, which contributes to climate change, could constitute a wrongful harm on conventional moral theories. In the face of similar examples, Young (2013) argues that adequately understanding the wrongs involved in structural and systemic harms requires significant rethinking of conventional normative frameworks—a project we shall return to in Chapter 9.

If states put autonomy first in this way, then it is unclear how they would be able to ensure that citizens are protected from diffuse and systemic harms (Dawson and Verweij 2008; Jennings 2009).[3]

5.2 Rethinking Autonomy

Much work within public health ethics has stayed broadly within the confines of the 'autonomy first' approach, but has sought to expand what is allowed by it, through reinterpreting both the nature of autonomy, and the extent to which activities such as food choices meet a minimum threshold of autonomy. The underlying project has been to show that public health activity may not be so much in tension with a properly worked out 'autonomy first' approach.

Such contributions have struggled to establish a consensus on the correct interpretation of autonomy, because the idea of autonomy functions in at least two different ways in 'autonomy first' accounts. First, autonomy is used to qualify *choices*, the key claim being that autonomous choices are worthy of a respect that nonautonomous choices are not. Second, autonomy is used in the sense of 'infringing someone's autonomy', where autonomy refers to a sphere in which the individual is sovereign. While these two views are often combined (both in the popular imagination and in the philosophical literature), it is important to notice that they are distinct, and there is a degree of tension between them (Wilson 2007a). Autonomy as sovereignty assumes that, so long as a self-regarding choice is genuinely attributable to an individual, there is a duty on the part of the state not to interfere with it, regardless of whether that choice has been well-deliberated or is in line with the chooser's deepest values. Autonomy as autonomous choice sets a rather more demanding standard: on its view, only choices that meet the appropriate standard of autonomy in deliberation are to be immune from state interference.

Many individuals make use of autonomy as sovereignty in some contexts and autonomy as autonomous choices in others. Health professionals frequently hold that patients with mental capacity should have the right to refuse treatments for any reason or for none (autonomy as sovereignty), while also holding that food choices are not very autonomous and so the state should play a much greater role in shaping diets by preventing the sale of oversize sodas (autonomy as autonomous choice).

---

[3] It is, of course, possible for individuals or groups of citizens to organize non-state means of protecting themselves against some such threats (for example by setting up a slimming club). But in cases such as obesity where there are a number of systemic drivers, it is likely that such individual measures will prove ineffective.

This looks potentially inconsistent. One way for the 'autonomy first' theorist to respond would be to impose the same standard everywhere for what is required by respect for autonomy, settling on an autonomy as sovereignty model or on an autonomy as autonomous choices model. Given the scale of the differences between approaches adopted in clinical medicine (where the usual assumption is that patients should have the right to refuse any intervention), and government tax policy (where it is assumed that governments have a right to enforce payment regardless of objections individuals may have) a 'one-size-fits-all' approach would have radical and counterintuitive implications.[4]

An alternative is to adopt a contextual approach to autonomy: on such a view, respect for autonomy may still come first, but what respect for autonomy requires will differ according to normatively relevant features of the situation.[5] Following this line of reasoning, it would be open to the 'autonomy first' theorist to develop a normatively rich account of context that allowed nuanced and satisfying discriminations to be made about what respect for autonomy requires in different circumstances.[6]

It might be that a properly specified conception of respect for autonomy would allow public health measures that might initially seem to be ruled out by an 'autonomy first' approach, such as where measures undertaken are necessary to restore and ensure 'deep autonomy', or have a strong democratic mandate from the population to whom they apply, or are necessary to supply health-related public goods, or to overcome asymmetric information (Nys 2008; Anomaly 2011; Wilson 2011a).

Notwithstanding this possibility, it is argued that even such more nuanced 'autonomy first' approaches are still significantly out of step with much of actual public health practice, which frequently involves interferences to improve population health without any attempt to show that the choices interfered with are compromised in terms of their autonomy, or infringe the sovereign domains of others (Dawson 2011). Health and safety inspections of restaurants, compulsory seatbelts in cars, product safety standards, and water fluoridation provide good examples. Recent years have seen running battles between those who assume that the 'autonomy first' approach is basically sound (and so much the worse for

---

[4] See, for example, Flanigan (2013), who argues that the doctrine of informed consent (and its corollary power to decline interventions for any reason or for none), which governs clinical interactions, should also apply to public health policy.

[5] Elizabeth Anderson suggests such a view when she argues that 'to respect a customer is to respect her privacy by not probing more deeply into her reasons for wanting a commodity than is required to satisfy her want. The seller does not question her tastes. But to respect a fellow citizen is to take her reasons for advocating a position seriously. It is to consult her judgment about political matters, to respond to it in a public forum, and to accept it if one finds her judgment superior to others' (Anderson 1993: 159).

[6] Two accounts that show how such a contextualist approach to autonomy might be constructed (though neither supports an 'autonomy first' approach to public health) are Manson and O'Neill (2007) and Nissenbaum (2009).

public health practice) and those who assume that public health practice is basically sound (and so much the worse for the 'autonomy first' approach).

The next two chapters aim to reconfigure this debate. This chapter performs a negative role, in examining the worry that interventionist public health policies are problematic because they are paternalistic. Chapter 6 makes a positive case that there is a right to public health and that states violate their citizens' rights unless they take positive action to systematically reduce risks to health. This chapter makes two main arguments. First, it is a mistake to attempt to transplant claims about the wrongness of paternalism in the doctor–patient relationship into claims about the wrongness of paternalism in public health policy. Whether an intervention should count as paternalistic depends both on the goals of the intervention, and on whether those affected by it consent to it. Both of these elements are much more complex and ambiguous in the case of policy interventions than in one-to-one cases. It will typically be the case that at most some (but not all) of the motivations and justifications for an interventionist policy are paternalistic, and only some (but not all) citizens will dissent from any given interventionist policy. So it is much more difficult to make sense of the claim that a given policy is paternalistic than is usually thought, and relatively few public health policies can confidently be labelled as paternalistic. In addition, two of the elements that make paternalism problematic at an individual level—interference with liberty and lack of individual consent—are endemic to public policy contexts in general and so cannot be used to support the claim that paternalism in particular is wrong.

Second, even where paternalistically justified policies can be identified, the arguments against them are less strong than is often thought. Arguments against paternalism come in two main types: (a) antipaternalist arguments which claim that avoiding self-regarding harm which is adequately voluntary never provides a reason in favour of a policy regardless of the magnitude of the harms it avoids and (b) arguments that policies which interfere with people's autonomous choices will very rarely if ever be of net benefit to those whose lives are interfered with.

Nonpaternalistic interference with liberty presents a dilemma for antipaternalists who would object to interventionist public health policies. Either the antipaternalist must hold that it is never legitimate to interfere with personal sovereignty for nonpaternalistic reasons, or that it is sometimes legitimate. If nonpaternalistic interference is never legitimate, then any single individual who would suffer a minor infraction of liberty would then have a veto against government policy. Governing in the public interest would become impossible. If, however, nonpaternalistic interference in an individual's sovereign zone is sometimes legitimate, then antipaternalism—perhaps surprisingly—may do little to rule out interventionist public health policies.

Policies do not need to benefit all individuals in order to be ethically justifiable. Policies such as redistributive taxation may be justifiable notwithstanding the fact that they leave richer citizens worse off. So objections to interventionist public health policies that focus on the balance of harms and benefits to specific individuals are much more convincing in cases of genuinely paternalistic policies (which aim to benefit individuals without their consent), than in the much more common case in public health where a policy infringes the liberty of some in order to bring benefits to others. In the case of cases that are genuinely paternalistic, what matters most is whether the infractions of liberty and autonomy are proportional to the health benefits to be achieved, and whether the measures to do so can command public confidence.

5.3 Paternalism, Coercion, and Government Action

The definition of paternalism has been the subject of a large literature, with at times some subtle cases being traded back and forth.[7] The points made in this chapter do not turn on any of these controversial edge cases, and I shall assume that the definition put forward by Dworkin (2020) is broadly correct. Paternalism, on this account, has three features: first, it involves an interference with either the liberty or autonomy of the person subjected to the paternalism.[8] Second, the interference is done without the consent of the person interfered with. Third, the interference is undertaken in order to benefit the person interfered with.[9]

It is standard to distinguish between soft and hard paternalism. Soft paternalism involves interference with a person's choices, where those choices are reasonably believed to be less than adequately voluntary (for example, interfering with an addict's ability to get hold of their drug of choice). Hard paternalism involves interference with choices, where the person interfering has no reason to think that these choices are less than adequately voluntary: for example, preventing someone from taking their own life, even if they have thought long and hard about the decision and their choice is made autonomously.

The soft/hard paternalism distinction refers to the voluntariness of the choices interfered with, and is a separate question from the coerciveness or otherwise of

---

[7] Influential attempts at a definition include Dworkin (1972); Gert and Culver (1976); Feinberg (1986); Shiffrin (2000).

[8] I use liberty to refer to absence of external constraints to action, and autonomy to refer to an individual's ability or right to make choices in line with their values. An addict may lack autonomy, even though they do not have any external constraints on their action; and a resourceful and resilient prisoner may still make autonomous decisions despite their deprivation of liberty.

[9] 'I suggest the following conditions as an analysis of X *acts paternalistically towards Y by doing (omitting) Z*: 1. Z (or its omission) interferes with the liberty or autonomy of Y. 2. X does so without the consent of Y. 3. X does so just because Z will improve the welfare of Y (where this includes preventing his welfare from diminishing), or in some way promote the interests, values, or good of Y' (Dworkin 2020). Notice that there are some such as Shiffrin (2000: 216) who define paternalism in a broader way to encompass taking over or controlling 'what is properly within the agent's own legitimate domain of judgment or action', whether or not this is done for that person's benefit.

the means employed to interfere with these choices.[10] So it is possible to interfere in a soft paternalistic but very coercive way, or a hard paternalistic but non-coercive way. An example of very coercive soft paternalism would be arresting anyone who has been diagnosed as a problem gambler if they set foot in a casino. Examples of noncoercive hard paternalism are slightly harder to come by, but might include providing someone who endorses their current identity as a smoker with a very large cash reward if they quit.[11] Our main interest in this chapter is in the justifiability of interference with autonomous choices. If, as will be argued, interference with autonomous choices is often legitimate in public health policy, interference with non-autonomous choices be justifiable a fortiori.

Paternalism is defined by its aim, not by its consequences. Paternalism requires that the interference with choice be done in order to benefit the recipient. It is not acts of interference with choice on their own that are paternalistic, but acts of interference in conjunction with the end of benefiting the persons interfered with. It is essential for paternalism that the interference *aims* to benefit the person interfered with, not that it succeeds in so doing. Interference that aims to benefit someone, but in fact leaves them worse off, could still count as paternalism.

The net result is that a very similar looking interference with someone's liberty could be either paternalistic or nonpaternalistic depending on its goal. If I stop a child from eating all the cake because I am worried that he will be sick, that would be paternalism. If I stop him from eating all the cake only because I want to ensure that there is enough left for other guests, that would be not be paternalism. In this case, child may resent the intervention equally regardless of whether it is paternalistic, but there are other contexts in which it is thought that paternalism implies a particular kind of disrespect that is absent from other kinds of restrictions on liberty. Darwall gives a useful example in which parents give their 40-year-old daughter a hard time about not eating enough green vegetables, concluding that:

> The objectionable character of paternalism of this sort is not primarily that those who seek to benefit us against our wishes are likely to be wrong about what really benefits us. It is not simply misdirected care or even negligently misdirected care. It is, rather, primarily a failure of respect, a failure to recognize

---

[10] It is worth noting that lawyers and economists sometimes use the soft/hard paternalism distinction differently, to distinguish between coercive and noncoercive interferences. Where I refer to this latter distinction, I draw it simply in terms of coercive and noncoercive interventions.

[11] Questions of this kind have been considered within the literature on research ethics under the heading of undue inducement, and one controversy has been whether getting someone to take part in a research project that they would otherwise not have done by offering them a large amount of money undermines the voluntariness of their decision, or whether it should be viewed as altering the nature of the choice in such a way that is becomes rational to take the money. See Wilkinson and Moore (1997, 1999) and Emanuel (2004).

the authority that persons have to demand, within certain limits, that they be allowed to make their own choices for themselves.   (Darwall 2006: 267–8)

One important question is whether, and if so how, this moral insight transfers into the realm of public policy. Much discussion of paternalism in the philosophical literature focuses on simple cases where one individual acts paternalistically towards another. In such cases, a central feature is that the person who is treated paternalistically has preferences and values that are either reasonably clear to the paternalizer, or could easily be clarified by asking them. And so these are cases in which it would be relatively easy for the paternalizer to tailor their actions to what the paternalized person believes would benefit them. Suppose that a patient in a hospice makes clear that they want resuscitation to be attempted if they go into cardiac arrest, and they continue to desire this despite all indications that they would be very unlikely to receive any significant medical benefit from this procedure. It will usually be perfectly feasible for healthcare professionals to mark the patient for attempted resuscitation. Paternalism in this kind of context seems objectionable because it is easy to ask the individual what they want and to tailor the treatment to their preferences, but the paternalizer either does not bother to ask, or asks and then overrides the paternalized's preference. It is not hard to see why paternalism like this is thought to involve a wrongful disrespect to the person's status as a competent agent. In public policy contexts, the status of paternalism is much more complex, as the next section explores.

5.4 The Very Idea of Paternalistic Policies

Interferences with liberty are not per se paternalistic, but only in the light of an understanding that their aim is to benefit recipients against their will. Thus, the claim that a law or a public policy intervention is paternalistic presupposes an account of its goal. Asking what the goal of a public policy intervention is could be interpreted as a factual question about the psychological states of the legislators or officials who shaped it, or as a normative question about what the best normative justification would be for the policy in its current form. On neither reading is it straightforward to make sense of a policy or law as paternalistic.

If a psychological reading is adopted, then there will not usually be a single goal of a given policy intervention. Individual policymakers will have a variety of reasons for advancing a particular policy. For example, a minister may have several of the following reasons (among others) for introducing legislation to increase the rate of taxation on alcohol: wanting to reduce levels of violence in society, providing more tax revenue for a government, complying with a WHO recommendation, showing themselves to be capable of standing up to big business, as well as the paternalistic motivation of trying to benefit people without their consent by making it more expensive for them to drink. As a policy goes through various processes of debate and refinement, it will be altered to accommodate different interests and power groups, and in a functioning democracy the final result is likely to be somewhat different from the initial vision that animated it. Even after a policy is enacted, it will be differentially interpreted by government actors such as civil servants or police who have the duty to enforce it (Lipsky 2010).

If a normative reading of the goal of a public policy intervention is adopted, then there is a problem of circularity. If the goal of a public policy intervention depends on what the most plausible normative justification of the policy would be, then our judgements about the justifiability of paternalism will determine the types of policies we describe as paternalistic. As Peter de Marneffe puts it, 'if paternalistic justifications are illegitimate, as some antipaternalists surely believe, then the "best rationale" will never be paternalistic, and therefore no policy will ever be paternalistic according to this account' (de Marneffe 2006: 73).

In short, any real-world public policy intervention will have been formed by multiple psychological intentions, and any real-world public policy intervention will have multiple plausible normative justifications. Hence, both psychological and normative accounts need to provide an account of when either an intention of benefiting people without their consent, or a justifying reason of benefiting people without their consent, is sufficient to render the policy paternalistic. Two extreme views might be to say that (a) a policy is paternalistic if *any* of the motivations or justifications which explain its shape are paternalistic, or (b) a policy is paternalistic only if *all* the motivations or justifications which explain its shape are paternalistic.

Both of these extreme positions are obviously inadequate: (a) would allow that a particular policy is paternalistic even if paternalistic intentions or justifications played only a very small part in its genesis and shape. Option (b) would in practice mean that no policy would be found to be paternalistic, as there will always be a possible nonpaternalistic justification or motivation for policies which have a large paternalistic commitment. However, it is far from clear that it would be sensible to say something along the lines of 'a policy is paternalistic if *most* of the motivation behind it is paternalistic', or 'a policy is paternalistic if its main normative justification is paternalistic' given (a) the great difficulties involved in counting psychological motivations, and (b) the fact that our interpretation of what the main normative justification for a law is will be influenced heavily by our prior normative commitments.

De Marneffe argues that given these difficulties, we should adopt an account of paternalistic policies that combines both psychological and justificatory elements. On this account, it is a necessary condition for a policy's being paternalistic towards A that the government 'has this policy only because those in the relevant political process believe or once believed that this policy will benefit A in some way' and that the policy 'cannot be fully justified without counting its benefits to A in its favor' (de Marneffe 2006: 73–4). However, as de Marneffe admits, even this approach is not wholly satisfactory: the definition may be both too narrow and too broad. For example, it may be too narrow, in as much as it would entail that a policy which is motivated by avowedly paternalistic intentions would not count as paternalistic, if a sufficient *non*paternalistic justification of the policy should be available. And it may be broad, in as much as it classes a policy as paternalistic whenever paternalistic motivations and justifications play a *necessary* role in it. But it will often be the case that nonpaternalistic motivations and justifications such as harm reduction also play a necessary role in the genesis and justification of a policy which is rightly described as paternalistic according to de Marneffe's criteria. So if we define paternalistic and nonpaternalistic in parallel ways, then such policies would simultaneously be paternalistic and nonpaternalistic.[12]

Opponents of paternalism in public policy thus must overcome significant difficulties even to state what it is that they are objecting to. I shall assume that, following Husak (2003), in so far as there is a cogent case against paternalism in public policy, it must be a case against types of *justification* for policies, rather than against policies per se. If this is the case then, as Husak argues, an antipaternalist should not say of any particular law or policy that it is wrong because it is paternalistic, but rather make the more restricted claim that the policy or law 'is unjustified in so far as it exists for paternalistic reasons' (Husak 2003: 391). This would leave open the possibility that while the policy or law was unjustified in so as it exists for paternalistic reasons, there remains an adequate nonpaternalistic justification for it.

5.5 The Unavoidable Coerciveness of States

Paternalism has three elements: interference, lack of consent, and aiming to benefit. We have seen how difficult it is to make good on the claim that a particular public policy *is* paternalistic. The next point is that many public policy interventions unavoidably share significant features of what makes paternalism problematic at an individual level, whether or not the policy aims to benefit citizens without their consent.

There are inherent limits to how well public policy can be calibrated to individuals' values and choices. Some policies are impossible to exclude people from, such as those that provide public goods such as clean air or a national defence

---

[12] De Marneffe could instead define nonpaternalistic in such a way that it encompasses all and only those policies which are not paternalistic in his sense. But without further explanation this would be an odd move: if both paternalistic and nonpaternalistic reasons would be necessary to justify a given policy adequately, why then conclude that we should categorize the policy solely as a paternalistic one?

policy. In other cases, it would be possible but prohibitively expensive to tailor a policy to individuals' choices (such as accommodating those who object to water fluoridation by building separate non-fluoridated water pipes to the houses of the dissenters). Even where it would be possible to grant individualized exceptions to a policy, it may be unfair or self-defeating to do so. Maintaining goods such as potable water or residential amenity requires the great majority of citizens and businesses to show consideration for others. Allowing some to obtain the benefits of the policy without contributing to the sacrifices that the policy requires would amount to an official sanctioning of free-riding. Such policies may also be self-defeating, as many citizens are willing to moderate their behaviour to maintain communal goods only if they believe that others are similarly motivated.[13]

More generally, it is simply not feasible to gain individualized consent from all citizens affected by a government policy. Not only would it be too expensive and too burdensome on the electorate to get each to vote every time a minor rule or policy were changed, but even if a government could consult each affected person and get their consent or dissent, it is unclear how it should interpret the results of any such referendum. If governments took a strict analogy with the case of individual medical treatment, they would have to offer each citizen an individualized veto, and say that even one dissent would be enough to render the policy a no-go. If governments were to take individualized consent this seriously, policymaking would be completely stymied. They could not achieve unanimity *either* for keeping the status quo, *or* any proposed reform to the status quo.[14] For any proposed policy, at least some would object, and hence it would turn out that *no* government policies were legitimate.

Public policy is a fairly blunt instrument, and governments will inevitably interfere in their citizens' lives in a myriad of ways, forcing them to acquiesce in public policy interventions whether or not they individually consent to do so. Even those elements of what a government does which merely offer options (such as a free healthcare service which people can attend if they want to) are paid for through taxation which is coercively extracted. So two elements of what is found problematic about paternalism in an individual healthcare context—interference and lack of individual consent—are endemic to public policy more generally.

Not everyone would agree that this type of state interference is legitimate—even in a well-run democracy.[15] However, my aim in this chapter is not to defend the ethics of interventionist public health policy to the satisfaction of those who

---

[13] Chapter 10.6 looks at the ethics of allowing conscientious objections to mandatory vaccination policies in more detail.

[14] As O'Neill (2002a: 163) puts it, 'Neither the status quo, nor any single route away from it, is likely to receive consent from all: unanimous consent to public policies is unachievable in real world situations.'

[15] For a well-known view which draws the conclusion of the *illegitimacy* of state-sponsored coercion, see Wolff (1970).

do not think that the state ever has legitimately coercive power. Rather, it is to examine the cogency of positions that do not object to all exercises of state power, but do object to interventionist public health policies on the grounds that they are paternalistic. Given that nonpaternalistic policies may be coercive and infringe liberty to exactly the same extent as paternalistic policies, it follows that those who want to defend the wrongness of justifying policies paternalistically need to show that there is something wrong about the paternalistically justified policy over and above its infringement of liberty.

5.6 Against Antipaternalism

The literature reveals two basic lines of argument against paternalistically justified policies. On the first line of argument (exemplified by Joel Feinberg), which I shall call *antipaternalism*, it is never a reason in favour of a policy that it is probably necessary to prevent self-regarding harm which is adequately voluntary.[16] Antipaternalism thus refuses to put avoidance of self-regarding harm into the balance when we are considering which policies to adopt.

The second line of argument does not rule out reduction of self-regarding harm as counting in favour of a policy in principle, but argues that for reasons connected either to the limits of government competence, or to the nature of the human good, governments will do worse in promoting the well-being of their citizens if they adopt policies which are justified on paternalistic grounds than if they refrain from so doing. In a nutshell, on the second line of argument paternalism does not work. As we shall see, neither line of argument provides a convincing general objection to public health policies.

Antipaternalists argue that the salient moral difference between morally permissible public policies—which will unavoidably interfere with the lives of some individuals without their individual consent—and morally impermissible paternalism is that the interference in the case of paternalism wrongfully trespasses on areas of the individual's life where the individual should be sovereign. Antipaternalism involves two kinds of claims: first, a negative claim that it does not count in favour of a policy if it is probably necessary to prevent self-regarding harm which is adequately voluntary. Second, a positive claim that so doing is positively wrong, because it involves a kind of disrespect or wrongful usurpation of the decision-making authority of the person:

> By and large, a person will be better able to achieve his own good by making his own decisions, but even when the opposite is true, others may not intervene, for autonomy is even more important than personal well-being. The life that a

---

[16] I borrow this way of formulating antipaternalism from Shafer-Landau (2005).

person threatens by his own rashness is after all *his* life; it *belongs* to him and to no one else. For that reason alone, he must be the one to decide—for better or worse—what is to be done with it in that private realm where the interests of others are not directly involved.   (Feinberg 1986: 59)

Antipaternalism in the first instance is a claim about individuals' rights to make their own decisions even when that may be catastrophic for them. For example, the antipaternalist is likely to think that individuals have a right to smoke tobacco (without harming others) if they so wish. This is sometimes taken to have the extended implication that it would be wrong to curtail the actions of tobacco companies in trying to sell their harmful products but, as Hansson argues, this does not follow: 'a person's right to harm him or herself does not necessarily imply the right for others to facilitate or contribute to their self-harming activity' (Hansson 2005: 97). For example, if it is an individual's right to commit suicide and if on reflection they decide that they do not wish to continue living, it would not follow that a government acts wrongly if it, for example, limits the size of packs in which drugs such as paracetamol are sold to reduce suicide risk.[17]

Feinberg takes his antipaternalism to be a position about the *criminal* law. For example, he indicates that he is not against paternalistically justified taxation on smoking: taxing an activity, he argues, is both less coercive and less morally condemnatory than criminalizing those who perform it, and so does not automatically fall into the category of wrongful interferences in an individual's sovereign zone.[18] Feinberg thus would not have an in principle objection to hard paternalistic justifications for policies which *did not* criminalize self-harming behaviour. Presumably he would suggest (as I would) that when considering such policies the relevant question is whether the interference with liberty required to reduce self-regarding harm in a particular case is proportional to the good done.

Given that criminalization of individuals' behaviour in those individuals' best interests plays only a very small role the public health policies recommended by public health practitioners, Feinbergian antipaternalism is eminently compatible with most robustly interventionist policies. Hence, if antipaternalists wish to claim that paternalistically justified policies are wrongful even where they are not implemented via the criminal law, this would make them more radical antipaternalists than Feinberg himself was. It is this wider view that I shall concentrate on—though given Feinberg's preeminent position as the theorist of antipaternalism, I shall continue to draw heavily on his work.

Antipaternalists are committed to the claim that personal sovereignty *always* takes precedence over the reduction of self-regarding harm even when the harm

---

[17] For the effects the legislation to do this had in the UK, see, for example, Hawton et al. (2004).

[18] 'I object to criminalization of smoking because it is supported only by a paternalistic liberty-limiting principle that I find invalid, but I do not oppose taxing end cigarette use, even though it too is coercive in a proper sense, and its rationale would be equally paternalistic' (Feinberg 1984: 23).

to be avoided is very large, and the interference trivial. As Feinberg puts it, 'sovereignty is an all or nothing concept; one is entitled to absolute control of whatever is within one's domain however trivial it may be' (Feinberg 1986: 55). The negative claim of the antipaternalist position is that hard paternalistic reasons do not count in favour of a public policy. This leaves under-described the situation that governments face in cases such as water fluoridation, where most agree with the policy and think that it is a sensible way of improving dental health, but some find the policy objectionably paternalistic, in that it interferes with their self-regarding behaviour for their benefit without their consent.

One way of interpreting the antipaternalist position would be to take Feinberg's claim about 'absolute control' within one's domain literally, and to interpret it as saying that the individual should have an absolute veto against any policy which would end up with behaviour in their sovereign domain being interfered with. On such a view, if even a *single* person objects to a policy's interference into their sovereign zone, then the policy treats them wrongfully, and so should not be implemented. This interpretation of antipaternalism would allow a single person to stymie the rest of society's attempts to regulate their own behaviour. This might seem disproportional, and arguably incompatible with the basic idea of personal sovereignty. One important element of sovereignty is the right to bind the will by making agreements with others that will curtail liberty, and so it might be just as disrespectful to persons' moral standing to deny them the authority to do so as it would be to interfere with their private behaviour for their own benefit.

Feinberg denied that the zone of personal sovereign control should be interpreted in this absolutist way. He argued that a policy to which the majority agrees and a minority object will not usually count as paternalistic in his sense:

> When most of the people subject to a coercive rule approve of the rule, and it is legislated (interpreted, applied by courts, defended in argument, understood to function) *for their sakes*, and not for the purpose of imposing safety or prudence on the unwilling minority ('against their will'), then the rationale of the rule is *not* paternalistic. In that case we can attribute to it as its 'purpose' the *enablement* of the majority to achieve a collective good, and not, except incidentally as an unintended byproduct, the enforcement of prudence on the minority.
>
> (Feinberg 1986: 20)

Feinberg presupposes that one's zone of personal sovereignty prevents one from being subjected to paternalistic interference, but it does not prevent one from being subjected to an interference that would have the same effect on one's ability to lead one's life, but is aimed to allow the majority to achieve a collective good. This revelation threatens to completely undermine the ringing claims about the importance of personal sovereignty. As Grill (2009: 149) argues, Feinberg does not appear to consider 'the restriction of the options of the minority to be in itself a moral obstacle to enactment of policy once the majority has consented'. Thus, Feinberg's position amounts to 'accept[ing] that societies with majorities bent on zealous self-regulation may impose strict health regimes on all citizens' (Grill 2009: 149). This is, to say the least, a rather odd result for a position which is supposed to be archetypally liberal. However, I don't think that Feinberg has made a simple mistake. These difficulties in finding a coherent and plausible response to a range of cases in public policy are inherent to absolutist versions of antipaternalism.

Any position which claims that hard paternalism is never justified, but allows equal interferences with liberty when they are justified by reasons other than paternalistic ones, faces deep challenges. We saw earlier that it is very difficult to make good on the claim that any particular public policy *is* paternalistic. Any attempt to do the ethics of public policy that depends on being able to draw clear distinctions between paternalistic and non-paternalistic policies faces not just theoretical problems but also practical ones, as it seems that nearly all public health policies that are criticized as paternalistic could also be described in such a way that the interference with the liberties of a minority is an unfortunate side effect of the pursuit of the common good.

The better approach to the philosophy of public policy is not to pick out paternalism as a particular evil to be avoided, but instead to conceptualize unwanted intrusions into choices and liberty as the basic category. The question then becomes one of when it is proportionate to enact public health policies that some will object to. Relevant factors in considering this are how significant the choices are which are interfered with, how coercive the means of interference are, how great the benefit is that the policy is expected to lead to and how these benefits will be distributed, and what proportion of people object.

5.7 Justifying Public Health Policies to which a Minority Object

There are various ways in which a policy can fail to benefit its recipients. Most straightforwardly, the policy might be framed in a way which fails to take account of important systemic effects, and due to unanticipated interactions with other policies, leaves its recipients worse off. As Part I explored, this is a general problem which applies to all public policies—whether or not they involve interferences with liberty in order to improve population health. An effective argument against interventionist public health policies would need to show that there are reasons for thinking that interventionist public health policies are either more likely to fail to achieve their ends than other policies, or more likely to have undesirable ends. Likewise, an effective argument against public health policies that appear to be paternalistically justified would also have to be built on arguments specific to this case.

Will interventionist public policies that appear to be paternalistically justified do more harm than good? Mill certainly thought that policies that intervened with the force of the law where someone's conduct would not cause harm to others would be likely to do so: 'But the strongest of all the arguments against the interference of the public with purely personal conduct, is that when it does interfere, the odds are that it interferes wrongly, and in the wrong place' (Mill 1977 [1860]: 283). Mill bases his argument here on the importance of individuality. To the extent that a policy is paternalistically justified, it interposes the judgements of public officials between an individual and their own conception of their good. It is difficult for governments to benefit citizens by assigning what seem (from the government's perspective) to be sensible weights to values, given that what makes each individual's life go well depends on features unique to that individual. Doing so will be likely to lead to rankings of values which make individual citizens' lives go worse. Moreover, the best life is a self-chosen one in which one develops one's own particular talents and inclinations, rather than having a life-plan imposed. Even if the weighting of values implicit in an interventionist policy is *better* than the one the person would have chosen for themselves, the fact that this weighting of values is *imposed* nonetheless makes the life that contains this superior weighting of values less good than the life with a freely chosen inferior weighting of values.

Mill made this as an argument against paternalistically or moralistically justified policies, rather than against public policy intervention more generally. To the extent that a policy is justified on non-paternalistic grounds, such as by the attempt to reduce health inequalities, or to prevent systemic harms, it involves weighing the interests of those whose lives are interfered with against the interests of others. There is little reason to think that individuals have a particular expertise in fairly weighing their own interests against those of others. Indeed, the widespread requirements for declarations of conflict of interest and for individuals to stand aside in official decision-making when their particular interests are engaged, suggests the opposite. To the extent that public health policy aims to redistribute risks between individuals, whether such policies are fair and justifiable, needs to be deliberated from a perspective that takes account of all relevant interests. For example, a policy to promote active transport through cycling, which involves creating many miles of additional cycle lanes, and correlatively reducing the road width available to motorists, may make the lives of dedicated motorists a little worse, but the policy may well be justified all things considered. The fact that some citizens object (even violently) to a public health policy should not be treated as a decisive objection to it.

Despite the fact that it will be difficult to disentangle paternalistically from non-paternalistically justified policies in practice, it is worth focusing for the sake of the argument on paternalistically justified policies, in order to clarify the extent to which Mill is right to worry that they are likely to do more harm than good.

Mill's argument has two presuppositions: first, the empirical claim that values differ between people in ways that renders government attempts to help epistemically problematic; and second, the normative claim that people's individual valuations are a better guide to what would make their life go well than their government's judgements. Both are contestable assumptions in the case of health. It is useful to distinguish between, on the one hand, controversial values and, on the other, controversial rankings of values.

Chapter 6 will make the case that there is a right to health, and that this implies a right to public health. To briefly anticipate this argument, health has a value as an all-purpose enabler for any number of other ends we might value for their own sake. Given this, even if someone places no particular value on their health per se, it will very often be the case that health enables them to pursue the goals they do value even better. If this is correct, then it is reasonable for governments to take health to be an uncontroversial value to be promoted; or if governments cannot assume that goods like health are of value then there is little if anything that governments can legitimately do.

While the relative weightings of health as against other goods such as liberty will be controversial, it is not feasible for governments to avoid taking stances which incorporate controversial rankings of health against other goods. For example, it is simplistic to imagine that governments will do better in helping their citizens to achieve a good life if they adopt a laissez-faire approach. The idea that a laissez-faire approach would allow each citizen to decide for themselves the relative weight that *they* want health to have as compared to other goods in their lives is illusory. This is because individual trade-offs about the value of health relative to other goods are made in the context of broader choices about the structure of society. Many choices about one's health (such as to be able to get exercise by walking to and from work) are only feasible if a whole set of background conditions, like street lighting, maintained sidewalks, and proper town planning are in place. So, if a government responds to value controversy by adopting a laissez-faire attitude, this will by no means allow all to balance health against other goods in a way they deem optimal. Rather, it will favour some rather than other forms of life and some rather than other sets of choices. As Chapters 6 and 8 discuss, attempts to evade this responsibility by adopting a policy of state inaction are likely to violate individuals' rights.

Policies that aim to promote health by largely or completely co-opting a person's will for their own benefit, and back this up with the coercive force of the criminal law, certainly appear to be difficult to reconcile with the importance of individual autonomy, and so might be taken to do more harm than good.[19] However, it is important to notice that there are a number of obvious instances in

---

[19] There are no such valid general concerns about the use of state coercion in cases of the criminal law where it is to prevent harm to others, as in cases like the criminalization of interpersonal violence.

which the requirement to take precautions is legally enforced for what would appear to be cases of self-regarding actions, such as mandatory seatbelt and motorcycle helmet laws.[20] The apparent incompatibility between Mill's principled objections to criminalizing conduct that does not involve harm to others, and mandatory seatbelt laws, could be resolved either by calling into question Mill's principle or by resolving that mandatory seatbelt laws are unjustifiable (Flanigan 2017).[21]

While such laws may have taken some time to be enacted and to bed in, and in some countries were controversial during this period, it is notable that as the new norms became entrenched they came to be perceived as much less onerous to comply with. Once someone has become used to wearing a seatbelt, and a set of norms around so doing have been entrenched, then the costs to individuals of so doing are minimal. Most car drivers and passengers simply do not really notice this level of coercion, as it has become an automatic part of their driving routine (often these days they will be aided in remembering by automatic reminders if they have not buckled up).

Thus, for the vast majority, the fact that wearing a seatbelt is mandatory does not really impinge on their lives. While the choice to ride unbelted has been taken away (or at least made much more expensive by making the rider liable to a fine), for most this is not much of a loss. For most people, it does not constitute a prob- lematic attempt by the government to impose something that they would not otherwise have chosen. Being forced to wear a seatbelt does not threaten their ability to be the author of their lives, as what matters for self-authorship is to be able to make choices about the important decisions, and to have enough choices, rather than to have as many choices as possible about every single minor detail (Raz 1986).

While there remain some individuals unhappy at mandatory seatbelt policies, it would not be plausible to claim that seatbelts have overall done more harm than good. Even trenchant critics of paternalism such as Snowdon (2017) do not deny that seatbelts have saved lives, but argue that they are problematic because 'they

---

[20] Of course, there is also a wide range of circumstances in which state interference in self-regarding activity would be unethical, even if it would be an effective way of preventing individuals from harming their own interests. For example, in Edwards and Wilson (2012) we argue that the sig- nificance of choices in terminal illness may be sufficiently great as to make interfering with such choices illegitimate, even where the patient intends to make a decision to take part in research that looks decidedly unwise. Chapter 10.4 discusses the case of conscientious objection to vaccination, concluding that even in schemes that make vaccination mandatory, it should usually be allowed.

[21] A third alternative would be to argue for a non-paternalistic interpretation of such laws— arguing for example that they are to be interpreted as attempts by the majority to protect the common good. One instance in which this kind of reasoning is seen is when judges who are convinced that laws should not employ hard paternalism interpret laws requiring motorcyclists to wear a helmet as having an implicit harm-to-others justification such as avoiding extra medical care costs (Bayer and Moreno 1986). As my aim in the main body of the text is to show that even if it is accepted that the policy is paternalistic, this does not necessarily show it to be ethically unjustifiable, I leave this third alternative on one side.

patently breach Mill's harm principle and change public perception about the objectives of criminal law' (Snowdon 2017: 40).

What about those who do find the imposition of mandatory seatbelts a heavy burden? Let us suppose that being forced to take precautions spoils someone's enjoyment in driving, and this, for them, is not a matter of mere annoyance but something that goes deep. It is helpful to distinguish two kinds of reasons for which a well-informed person might want to partake in activities that are risky to health. First, there are cases where the danger to life or health is an accepted but not an intended element of the activity. Eating of the Fugu pufferfish in Japanese cuisine provides a good example. The fish is extremely tasty, but if badly prepared can lead to death by poisoning. The preparation of Fugu is very tightly regulated, and precautions are taken to minimize the risk. Diners accept the residual risk to health as a price worth paying for the culinary experience. To the extent that what is valued is the taste of the fish, rather than the danger, it will be accompanied by a desire on the part of diners to reduce or eliminate these residual risks to the extent possible.

Second, there are cases where danger is sought for its own sake. In these cases, the activity would lose its point if precautions were taken that reduced the risk. For example, Russian roulette could easily be made much safer by ensuring that the gun were not loaded, but this would undermine the main purpose the activity, in which embracing the possibility of death is paramount. Some cases, such as free solo climbing and perhaps unbelted riding may contain elements of both, depending on the proclivities of individual practitioners.

The aim of increasing safety by reducing avoidable risks to health and well-being is at the heart of public health activity, and so there is an obvious limit to how far public health policy should make space for the idea that self-endangerment is to be valued as an end in itself. Conversely, the point is well-taken that people often deeply and sincerely value activities that happen to be bad for their health, and will sometimes want to carry on doing those activities in the light of a lucid understanding of the attendant health risks.

This might initially seem like a zero-sum game, in which public health's gain is liberty's loss, and vice versa, but appearances are deceptive. As the complex systems approach to public policy set out in Part I should have made clear, there are multiple pathways to reducing health risks. How to reduce risks in ways that are less onerous for individuals is itself an important area of research. The importance of reducing these costs to individuals is one reason why Chapter 8 argues that where possible, health risks should be managed or eliminated by governments, rather than requiring citizens to take precautions on an individual basis.

So, what principles which should guide a state's policies to improve citizens' health, where these policies will involve restrictions on liberty or autonomy? First, it is obvious that the size of benefit to be gained or size of harm to be avoided by interfering matters. Other things being equal, the greater the expected benefit and the greater the expected harm to be avoided, the stronger the argument in favour of intervention. Second, the extent to which the population regulated endorses or consents to the policy matters. Other things being equal, the greater the percentage of the affected population who endorse an intervention (and the more enthusiastically they do so) the stronger the reason in favour of the policy. Third, autonomously chosen lives matter. Other things being equal, the more significant a choice is, the more important it is that a person has the opportunity to make a genuine or authentic choice and the more problematic it is to interfere with their choice. Fourth, liberty matters. Other things being equal, the more coercive a policy is, the more problematic it is.

Some cases of potential interventions for health will be clear-cut. Where there is a policy that will bring a great benefit, which is supported by the vast majority of people and involves only a mild interference with choices which are not generally thought to be significant, the intervention will be easy to justify. Where a policy will bring only a small benefit, and which is opposed by the vast majority of people and involves a coercive interference with significant choices, then the intervention will definitely not be justified.

The difficult cases will be those closer to the middle. Further normative work will help to clarify types of cases, but it is important to notice that we should not expect that abstract normative reasoning will be able to give us definitive once-and-for-all answers to these questions. This is because—if what I have argued is correct—which interferences are justifiable depends (among other things) on the level of general consent to the policy, and the significance of the choices interfered with. The level of consent will obviously vary relative to culture and time; and we will also have to take account of local differences in which choices are believed by particular communities to be significant. I take up these issues more concretely in the context of vaccination policy in Chapter 10.

5.8 Conclusion

This chapter has had two main themes. The first is that it is unhelpful to take the question 'is the policy paternalistic?' to be fundamental in the ethics of public health policy. There are severe problems in making good on the claim that a particular policy *is* paternalistic. Further, the marks of much of what makes paternalism morally problematic on an individual level—interference and lack of individual consent—are present in a wide range of public policy contexts, whether or not the policy in question can be described as paternalistic. It is better to frame the ethical question as about what types of justifications for public policies are legitimate and to frame the practical question for public policy as about which infringements of liberty are justifiable, without trying to stake too much on whether those infringements of liberty are paternalistic or not.

The second theme has been about how to specify at a general level which kinds of public health that infringe on liberty and/or autonomy are ethically justifiable. While there are myriad ways in which public policy can fail, public health policy is no more likely to fail than any other kind of policy. There are reasons to think that policies which aim to override individuals' conceptions of the good are more likely to fail to benefit individuals than those which do not, but it is important to remember that health has as strong a claim as any good to be an uncontroversial good for states to promote. Rankings of health relative to other goods will be controversial, but it is impossible for states to avoid taking controversial stances on the value of health. Policies that involve legally mandated precautions may seem to the vast majority of individuals to be justified notwithstanding the fact that they appear to be paternalistically justified.

The next chapter advances a positive case for interventionist public health by arguing that there is a right to public health.

6

The Right to Public Health

6.1 Introduction

This chapter develops a positive case for state action to protect and promote health as a duty that is owed to each individual. On this view, the state may violate individuals' rights if it fails to take cost-effective and proportionate measures to remove health threats from the environment. If this is correct, it is mistaken to approach public health as primarily a matter of individual entitlements versus the common good. Too little state intervention in the cause of improving population health can violate individuals' rights, just as too much can.

Chapter 5 introduced what I described as 'autonomy first' views, which take it for granted that it is difficult to justify state interference in the lives of competent adults unless the behaviours interfered with are compromised in terms of their autonomy, or would wrongfully infringe the autonomy of others. Such views focus on the main ethical duties as being ones of non-interference. Those under the sway of these ideas often critique state action to promote or protect people's health through the idea of the 'Nanny State', as if public health is by its nature likely to overreach and to become problematically paternalistic (Coggon 2018).

This chapter argues that we also need to guard against something arguably even worse: the *Neglectful State*. The Neglectful State is one that does not take easy steps that could have been taken to reduce risks to health, and as a result allows significant numbers to come to harm or death. The ethical challenge of public health policy is therefore not the one-sided one of avoiding Nannying, but the more complex task of steering a course between Nannying and Neglect. Once the full implications of a Neglectful State come into view, then it may be that it is more ethically dubious and more politically dangerous to err on the side of Neglect than of Nannying.

The right to public health approach developed in this chapter shares with traditional public health a concern for protection and promotion of health as a core concern of the good society. But it differs from it in that it explicitly frames protection and promotion of health as being grounded in the rights of individuals, and as something that should be undertaken for the sake of individuals, rather than as something that should be done for population benefit. On a traditional public health approach, there will be duties on the part of governments to tackle air pollution or obesity, but no corresponding entitlement on the part of citizens to insist that such measures are taken. The rights-based account places an individual entitlement to such public health measures centre stage.

The shift to an individualistic model also makes a significant difference to the content of public health obligations. As is well known, risk reductions at a population level may create expected health benefits for each individual that are so small that most individuals would prefer not to have the benefit if it came with any additional inconvenience at all. The right to public health account takes it as axiomatic that it is not sufficient to justify interferences solely on the grounds that they will improve population health, or even that they will increase overall well-being, but that it is necessary to do so via an account of what is justifiable to particular individuals. The ethical implications of this stance and what it entails for prioritising policies are assessed at length in Chapter 7.

The right to public health account transforms, but does not fundamentally contradict, the Millian paradigm (Dawson and Verweij 2008; Jennings 2009) that is so prevalent within bioethics writing on public health, and which was alluded to a number of times in Chapter 5. The Millian approach to public health begins from a suspicion about the overweening power of the state, and introduces the harm principle as a way of protecting individuals from this power:

> the only purpose for which power can be rightfully exercised over any member of a civilized community, against his will, is to prevent harm to others. His own good, either physical or moral, is not a sufficient warrant. He cannot rightfully be compelled to do or forbear because it will be better for him to do so, because it will make him happier, because, in the opinion of others, to do so would be wise, or even right. (Mill 1977 [1860]: 223)

Most modern commentators have followed Feinberg (1984) in thinking that the conception of harm invoked in the harm principle needs to be clarified by distinguishing between mere harms and wrongful harms, given that nearly all liberals assume that the mandate of state to intervene to prevent wrongful harms such as unconsented to violence is much stronger than its mandate to preventing non-wrongful harms such as those arising from fair competition.

Once the harm principle is rethought via a moralized conception of harm, the idea of rights violation becomes paramount. My account shares with Feinberg's harm principle the fundamental thought that protecting individuals against violation of their rights is a central justification for state activity. But it differs from it in arguing that Millians have been too narrow in their account of individuals' rights. Once the right to public health is introduced, and once it is clarified that this right can be violated by failures to reduce risk as well as by actual harms, and once the mechanism for determining when this right is violated by the state is clarified, then we have a robust account of public health that is in the spirit of liberalism.

This chapter makes an ethical rather than a legal argument, but it is worth noting that the existence of a human right to public health is well established within international law. The International Covenant on Economic, Social and Cultural Rights (ICESCR) (which has been ratified by 173 states) recognizes a 'right of everyone to the enjoyment of the highest attainable standard of physical and mental health' (ICESCR Art. 12.1), and explicitly requires states to undertake public health and health promotion activity (ICESCR Art. 12.2). States party to the covenant agree that they will progressively realize the highest attainable standard of health for their own citizens and take measures to better allow other signatories to do the same for their citizens (ICESCR Art. 2.1). It follows that states can violate the human rights of their citizens (and potentially citizens of other states) if they fail to take appropriate means to promote and protect health. Of course, the fact that a human right to public health is legally recognized does not by itself show that there is sufficient ethical reason to do so, though it might give some indication about how a moral right to public health might best be interpreted, and certainly helps us to ward off any scepticism that it would be impossible to implement such a right.

I begin by clarifying what it is to justify a rights claim. As will become apparent, even what would count as an adequate justification of a rights claim is contested. Thus, the ambitions of my argument are limited in one important sense. It is not directed to those who, for example, are sceptical about rights *tout court*, or who for theoretical reasons think that there cannot be positive rights, but rather to those who are willing to countenance that there are some rights that give citizens entitlements against the state, but are yet to be persuaded that these include a right to public health. I set out an argument why the relationship between health and human well-being makes it very plausible to think that there is a right to health, and why a right to public health follows from the right to health. I then argue that, just as in the case of the right to security, the right to public health justifies significant (though proportional) interferences with liberty.

If we take the right to public health seriously, then—just like other rights such as the right to security—this will justify the government in taking steps to protect everyone's rights. The right to public health account thus allows those who had previously adhered to the 'autonomy first' account a new way of understanding how robust public health policies are not only compatible with, but may be required by, their commitment to individual rights. The non-interference principle introduced in Chapter 5.1—namely that individuals have an entitlement that their adequately informed and adequately voluntary decisions not be interfered with unless the interference is necessary to prevent violation of rights or rectify existing rights violations—is a threat to the ethical justifiability of public health only on the assumption that there is no right to public health. If there is a right to public health, and this right entails that governments have a duty to take significant action to promote and protect health, then government action taken to avoid violating this right would by definition not count as interference that is ruled out by the non-interference principle. The non-interference principle is thus compatible with quite extensive government action, if this is necessary for the purpose of rights protection.

Thinking about public health in rights terms disrupts the traditional assumption that a different normative justification is required for health promotion (which attempts to promote population health by educating and empowering individuals) than for health protection (which aims to remove health threats from the environment). On my view, both are grounded in the right to public health. This chapter aims to defend the right to public health at a general level. Adequately specifying the content of this right for different contexts and in different areas of public health will take the rest of this book.

6.2 Justifying Rights Claims

The philosophy of rights is rife with disagreement both about what the function of a right is, and the role that rights in general should play in moral discourse, so I shall make a few remarks by way of clarification. My aim here is not to say anything new, but to map out the contours of agreement and disagreement within the literature (Wilson 2007b).

Rights, as I use the term, are high priority claims that are correlative to directed duties. To describe a duty as directed is to say that it is owed *to* an individual in such a way that if the duty is not performed appropriately, the person to whom the duty is owed has a privileged basis for complaint. In such circumstances, unless there are adequate excusing factors, the duty bearer wrongs the holder of the directed duty.

Directed duties are to be distinguished from non-directed duties, such as a general duty to maximize the good. If someone fails to perform a non-directed duty, this does not give any third party grounds for claiming that he or she has been wronged as an individual, even though there will be grounds for saying that the duty-bearer has acted wrongly (or at least sub-optimally). So an initial test for whether you think there is a right to public health is if you think it possible that someone could be wronged as an individual by government failure to take measures to protect health.

I shall assume that the duties correlative to rights must be of high priority, but apart from that shall not take a stance on what the strength of such duties must be.[1]

---

[1] The idea of a directed duty does not by itself entail that such a duty must be of high priority. It is possible to conceive of a directed duty that would easily be outweighed by other duties. However, it is commonly assumed that the claims correlative to such duties would not be rights. Suppose I make a commitment to you that I will come to your birthday party so long as I can swap my shift at work to be free that night. This generates a directed duty because, in the case where I do get the night off work, and then do not come to your party, you might legitimately mildly reproach me. But the strength of my duty to attend your party would fall far short of what would usually be thought to be required for a rights claim. (Rumbold 2018 argues that there is no need to think that rights by definition must entail duties of a high priority. As a conceptual claim about rights this is interesting, but given that I shall be

Directed duties correlative to rights can take various forms. There is no reason to think that each right will give rise to only one duty, or only one undifferentiated class of duty-holders. Indeed, the human rights literature (following an influential interpretation of Shue 1996 by the UN Special Rapporteur on the right to food Asbjørn Eide (see United Nations Centre for Human Rights 1989) assumes that human rights entail three separate sets of obligations: obligations to respect, obligations to protect, and obligations to fulfil. On this view, uncontroversial rights such as the right not to be tortured are commonly taken on closer analysis to involve not just duties on the part of individuals and states not to torture, but duties at a state level to prevent torture from occurring, and to bring to justice those who are suspected of torturing, as well as a duty to create systems in which torture is less likely to occur.

It is possible to conceive of a world without rights, albeit one that most of us would not find attractive (Feinberg 1970). But it is not the role of this chapter to make the case for rights-based thinking in general. Rather, I address myself to those who agree that there are some rights (perhaps to liberty, or not to be tortured), but are sceptical that these rights include a right to public health.

Will and interest theorists have waged a long and sometimes fractious dispute about the function of rights. Will theorists argue that the function of rights is to allow the right-holder to control the duties of others.[2] On this view, all rights (even rights not to be enslaved) can be waived, and only those who have the wherewithal to waive rights can be right-holders, ruling out young children and non-human animals as potential rights-holders. For the interest theorist, the function of rights is to protect or promote interests. Interest theorists argue that there is nothing incoherent about the idea of rights that the right-holder does not have the moral power to waive or annul, and so there is no bar to young children or non-human animals being right-holders.[3]

Cross-cutting the debate about the function of rights is another debate between intrinsic and instrumentalist conceptions of rights. Intrinsic conceptions of rights hold that rights are an intrinsic part of the furniture of the moral universe: human beings have rights because of the moral status that they already have prior to anything we do.[4] Rights instrumentalists think that rights

---

arguing that the duties correlative to the right to public health are of a high priority, it is not an argument that I need to engage with here.)

[2] In H. L. A. Hart's words, it views the right-holder as 'a small-scale sovereign' in the area in which the duty is owed, 'able to waive or extinguish the duty or leave it in existence' (Hart 1982: 183).

[3] This dispute partially explains the two conceptions of autonomy that are combined within the common discourse of respect for autonomy in health (see Chapter 5.2). In some contexts, autonomy is seen (in accordance with the interest theory of rights) as an interest in being able to take deliberated decisions about the shape of one's life. In other contexts, autonomy is seen as a feature that makes the individual a small-scale sovereign, able to refuse interventions for any reason or for none.

[4] As Warren Quinn (1993: 173) put it, 'It is not that we think it fitting to ascribe rights because we think it is a good thing that rights be respected. Rather we think respect for rights a good thing precisely because we think people actually have them—and...that they have them because it is fitting that they should.'

are constructed by us with a purpose of protecting, promoting, or making possible morally valuable states of affairs.[5]

What counts as a legitimate move in justifying a particular right depends on prior claims about the nature of moral justification. Styles of justifications of moral duties can either be foundationalist or holistic. In foundationalist justifications, moral duties are derived from moral principles that are more basic in the order of moral justification and explanation. These more basic principles must be supported by more basic principles, until we reach a small number of basic principles (perhaps just one) that are not themselves supported by any other principles.[6] Holistic approaches acknowledge that some duties can be justified by reference to others that are more basic in the order of justification, but they deny that this is always the case. On a holistic approach, justifications can also legitimately be a matter of mutual support of a variety of different concerns (as in reflective equilibrium).[7]

The upshot of all these disagreements about the function, fundamentalness, and justification of rights is that it not exactly clear what would count as giving a satisfactory justification of the right to public health (or indeed any other right). However, to the extent that this is a difficulty, it is also a difficulty for any views that believe in the existence of rights, and in particular those that aim to use rights talk to claim that public health activity is unethical. We can distinguish two levels of controversy about justifying particular rights: first, there are high level abstract questions about whether a right should be recognized at all. Second, there are a host of specificatory questions about what a given right should mean in practice: what its extension should be, who the duty-holders are, and how stringent the duties relative to the right should be.[8] My aim in this chapter is to defend the idea of a right to public health at a high level; a significant number of questions about duties and entitlements remain controversial and unresolved even if the broad idea of such a right is accepted. Chapter 7 extends this analysis by looking in depth at questions around how to prioritize which risks to health are the most urgent to address. Part III of the book examines a number of concrete cases in the light of systems theory and the right to public health.

---

[5] Instrumentalist justification can involve simple means–end reasoning (where a right is introduced solely as a means of reducing instances of a particular bad state of affairs), or constructions of a more complex and open-ended kind (such as where a right to privacy is entrenched because it is believed that giving individuals the guarantee that certain aspects of their lives will be undisturbed will allow them to pursue forms of life that are valuable in themselves and would not otherwise be possible).

[6] Will theorists tend to be keener to attempt to derive all rights from one single foundational right to freedom, whereas interest theorists are more relaxed about the possibility of the interests that are sufficient to ground rights being disparate, and thus there being multiple (and potentially conflicting) rights.

[7] Daniels (2018) gives a good overview of reflective equilibrium as an approach to ethical justification.

[8] For example, the right to life is widely accepted at an abstract level. Much more controversial is whether the best specification of the right to life includes an entitlement to the means necessary to sustain life, or if the right provides only protection against being unjustly killed. See, for example, Thomson (1971) on this question.

6.3 Arguing for the Right to Public Health

Is there sufficient moral reason to impose on governments those high priority directed duties that would be correlative to a right to public health? In what follows, I aim to establish a case for this right. My case relies in part on the positive effects that protection and promotion of health have for individuals' well-being, and in part negatively, on the wrong of those who have a duty to prevent others from being subjected to risks failing to do so.

It is difficult—even for those who are unsympathetic to the idea of a right to public health—to deny that health is of very great significance for well-being. Health is arguably a constituent part of well-being, but even leaving this on one side, health is an all-purpose enabler for any number of other ends we might value for their own sake. Even if an individual places no particular value on their health per se, it will very often be the case that health enables them to pursue the goals they do value. Individuals thus have a universal (or near universal) interest in their health. What is controversial is not the claim that health is of significant and near-universal importance for well-being, but the relationship between this claim and the further claim that governments have duties to protect and promote health, and that these duties are sufficiently stringent that individuals can sometimes legitimately claim that their right to public health has been violated.[9]

For the interest theorist, there is a relatively quick route to the claim that there is a right to health. Interest theorists such as Raz (1984) hold that a human interest which is (a) universal or near universal, and (b) of sufficient moral importance to well-being that it is legitimate to hold others to be under a duty to at least not frustrate this interest, and possibly help to promote it, is a sufficient ground for a right. Given this account of when an interest is sufficient to ground a rights claim, and the relatively uncontroversial status of the claim that individuals do have a significant interest in preserving their health, the outline case for the right to health is simple: health is not only a near-universal interest, but is also an interest of sufficient moral importance (given the relationship between health and flourishing) to hold some others to be under a duty not to frustrate this interest.

The main reason for thinking that the type of interest we have in our health generates a right to *public* health is that socially controllable factors make a very significant difference to rates of morbidity and mortality. Take road traffic accidents as an example. Government policies such as adequate speed restrictions, mandatory motorcycle helmets and car seatbelts, enforcement of proper seating restraints for children, and reduction of drink-driving make a very significant

---

[9] Thus Sreenivasan (2012), in arguing *against* a moral human right to health does not dispute that 'each human being has a morally very important interest in preserving (or restoring) his or her own health, in so far as this can be achieved through social action.'

difference to the risks citizens are exposed to (World Health Organization 2013).[10] Individuals cannot adequately control these risk factors for themselves: I may be able to reduce some risks by wearing a seatbelt and not myself speeding, but I cannot prevent drivers from speeding or drink-driving. These risks from others' driving behaviours are potentially catastrophic for my health, and can practicably be reduced by government action. Thus, through my morally important interest in my own health, I also have a morally important interest in risk factors being removed from the social environment.

Assigning duty-bearers can sometimes be complex, but in this case it is relatively straightforward: in order for an agent to be a duty-bearer under a right, it is sufficient that the agent has (1) the power and liberty to act and (2) a responsibility or duty of care in respect of the good in question (and so could sometimes do wrong in virtue of not acting). It should be uncontroversial that governments in functioning states do have the power to affect the social determinants of health; given the breadth and complexity of the types of causal relationships in play, governments are perhaps uniquely well suited to affect the distribution of social determinants of health. Whether governments have a responsibility to protect citizens' health should also be uncontroversial. Governments have a duty to act in the public interest. Acting in the public interest requires reducing significant and avoidable risks of harm to citizens where it is cost-effective and proportionate to do so.[11] Many risks to population health are significant and avoidable and can be proportionately and cost-effectively reduced. Therefore, there is a right to public health that citizens hold against their governments.[12]

Someone could object to this argument in a more or a less radical way. In a more radical way, they could argue that governments simply do not have a directed duty to reduce significant and avoidable risks of harm to citizens even

---

[10] Rates of death from road traffic accidents differ very significantly from less than five per 100,000 vehicles per year (Norway, Switzerland) to over 5,000 per 100,000 vehicles (Guinea, Burundi, Benin, Central African Republic) (World Health Organization 2013).

[11] Section 6.4 provides a more detailed analysis of when a public policy amounts to a violation of the right to public health. Until that point, the cost-effectiveness and proportionality constraints should be seen as placeholders. In broad terms, a proportionality constraint is necessary because it is not legitimate to violate other rights in the course of defending the right to public health. A cost-effectiveness constraint is necessary because new government activity needs to be considered in terms of its opportunity costs. Within a fixed budget, adopting a new intervention will often mean fewer resources available for other interventions; and so activities foregone need to be taken into account (Rumbold et al. 2016).

[12] This argument is supposed to establish that duties relative to the right to public health are held by governments, but it is not my intention to suggest that these duties fall exclusively on governments. While I do not argue for these duties or attempt to specify them in this chapter, I also think that there are duties on corporations, institutions, and individuals in respect of this right. I come back to these questions of how to distribute responsibility for public health activity in Chapter 8. I acknowledge that there are some cases—for example, failed states—where there is legitimate doubt about the power that a government has to affect the social determinants of health. However, this book focuses on public health policy within states that do have the capacity to mount an effective public policy if they want to. Failed states fall outside of the scope of the framework developed in this book.

where it is cost-effective and practicable to do so. I do not have a knockdown objection against such a view, but it is worth noticing that while it entails that there is no right to public health, it would also entail that there are no rights to safety or security either.

Less radically, someone could agree that governments have *some* duties to reduce risks of harm, but these justify at most a subset of traditional public health policies. First, someone could argue that even though governments do have a directed duty to reduce harm to citizens, health is the wrong type of interest to create such duties on the part of governments. For example, it might be objected that citizens usually care about their overall well-being, rather than their health per se, and that individuals often and reasonably make choices that involve trading off their health against other goods—as when someone chooses to be a soldier, or to climb a dangerous mountain. Second, it could be argued that while governments have a duty to reduce behaviour that imposes *uncompensated* risks on other people, there is no duty for governments to reduce behaviour that either does not impose risks on others at all, or where these risks are fully compensated.

Both objections raise valid concerns. There can be risks to health and well-being that are adequately consented to and which do not impose uncompensated risks on others; and it can sometimes be perfectly rational for individuals to endanger their health in order to pursue other goods that matter more to them. Where either of these conditions hold, it looks problematic for a government to *enforce* reduction of risks to health. But what results from these concessions is a requirement to better specify the appropriate targets of public health, and better specify what counts as a proportionate measure in reducing health risks, rather than a fundamental challenge to the right to public health. As was argued in Chapter 5, the proportionality of liberty-limiting public health measures depends on the normative significance of the choices interfered with, the extent to which the policy is coercive, as well as the degree of public support for such measures. Where an intervention would require disproportionate interference with liberty or autonomy, it should not be undertaken.

Thus, the claim made by the right to public health approach is not that public health must win out against claims such as those of autonomy, but the weaker one that there is *a* morally important interest in such health threats being removed from the environment, which is sufficient to ground a right on the part of individuals that the state take steps to do so. As in the case of all other rights, it will need to be interpreted in the light of other justified rights claims, and also broader social goals. The right to public health will not be infringed if there are countervailing normative commitments of even greater weight that prevent a state from taking steps to reduce a particular health threat. But the right might be violated if the government either directly increases health threats to individuals, or does nothing to reduce a health threat when it is practicable to do so.

6.4 The Right to Public Health as a Right to Risk Reduction

What would a right to public health look like in practice? The first thing to clarify is that no one will stay healthy forever, and so it is untenable to interpret the right to health as implying that there is a right to be healthy (even before we consider how to balance protection of health against competing interests). The right to health is thus best interpreted in terms of a right to the control and reduction of risk factors to health, and the availability of care for those who do become ill, rather than through any kind of unattainable goal of elimination of ill-health (UN Committee on Economic Social and Cultural Rights 2000: para. 8). The right to public health similarly is to be understood in terms of a right to risk reduction, rather than risk elimination. However, it would be a mistake to think that the right to public health could, or should, lead to something like a uniform reduction of all risks by, say, 30 per cent. Some risks to health—such as polio or asbestos in building materials—can practicably be completely eliminated, but others will be much more difficult to reduce.

The key specificatory question is thus: which risks need to be reduced and by how much if the right to public health is not to be violated? My answer in outline is the right to public health is best interpreted as requiring *accountability for risk reduction*.[13] That is to say, states need to have a systematic plan of how they will reduce health risks equitably over time. To the extent that they either fail to plan for such risk reductions, or fail to include within such plans risk-reducing measures that it would be practicable and cost-effective to implement, or fail to implement their plans, then there is a prima facie reason to think that citizens' rights have been violated.

There are some elements that should be included in any such plan—in particular those which the World Health Organization recognizes as being a 'best buy' in public health, where each such best buy is 'an intervention with compelling evidence for cost-effectiveness that is also feasible, low-cost and appropriate to implement within the constraints of the local health system' (World Health Organization 2011). These include childhood immunization, raising taxes on alcohol and tobacco, restricting access to retailed alcohol, enforcing bans on alcohol and tobacco advertising, and replacement of transfats with polyunsaturated fats across the supply chain. The presumption should be strong that the right to public health will be violated unless such best buys are not only a core part of public health plans, but are also actively secured through effective policies.

Other cases will be less clear-cut. Suppose an interested group of citizens mounts a campaign that the government should introduce a ban on the sale of

---

[13] This suggestion draws on Daniels's (2007) account of accountability for reasonableness, and particularly the account of the human right to health that Daniels gives in chapter 12 of his book.

sugary beverages in fast food restaurants above a certain size (say half a litre).[14] In doing so, they make the following argument:

1. Psychological research shows plausibly that portion size has a significant impact on the amount of food and drink that people consume. The larger the 'standard' portion, the more food and drink people will consume.
2. A culture of large portions (particularly of foods that are high in calories) is a significant driver of obesity.
3. Obesity is a very significant health threat.
4. A culture of large portions therefore presents an ongoing health threat.
5. If the government does nothing to counteract this health threat where proportionate and cost-effective means are available, it violates its citizens' rights.
6. Preventing the sale of excessively large sugary beverages is a cost-effective and proportionate means of reducing the health threat of large portion sizes.
7. Therefore, the government has a duty to reduce the sale of excessively large sugary beverages.[15]

Given the existence of the right to public health, arguments of this kind challenge governments either to act to reduce the health threat, or to explain why they do not plan to act despite the apparent availability of an easy way to improve population health. There are various things that the government officials could say in defending the lack of action. First, there needs to be a solid rationale for thinking that the intervention is sufficiently likely to be effective to be worth implementing. There is no duty to impose policies that are either proved not to work, or for which the evidence of effectiveness is very weak.[16] Second, even if the policy would plausibly work, it might not be suitable for implementation due to competing public health priorities.

At any one time, there are a very large number of potential policies that could reduce risks to health, and only a subset of these will be able to be implemented.

---

[14] This example is loosely inspired by the debates around New York City's soda size cap. Bateman-House et al. (2017) provide a detailed analysis of the actual policy.

[15] It is worth noting that an argument of this kind is implicit in the UN Committee on Economic, Social and Cultural Rights' General Comment 14: 'Violations of the obligation to protect follow from the failure of a State to take all necessary measures to safeguard persons within their jurisdiction from infringements of the right to health by third parties. This category includes such omissions as the failure to regulate the activities of individuals, groups or corporations so as to prevent them from violating the right to health of others; the failure to protect consumers and workers from practices detrimental to health, e.g. by employers and manufacturers of medicines or food; the failure to discourage production, marketing and consumption of tobacco, narcotics and other harmful substances...' (UN Committee on Economic Social and Cultural Rights 2000: para. 51).

[16] For the reasons discussed in Chapter 2.4, this is not to say that randomized controlled trials or systematic review evidence is required.

So it is vital that states have a transparent and reasonable system for setting priorities for public health risk reduction (Rumbold et al. 2016). This prioritization process need not, and probably should not, be a matter solely of multiplying the size of the risk reduction by the number of people affected. To the extent that an approach to public health is rights based, it needs to be justified with reference to reducing individual, and not merely population, risks. Moreover, factors such as the distribution of reductions in health risks also matter. Specifying in more detail what the contents of such plans ought to be requires detailed reflection both on which risks to health are most important to ameliorate (Chapter 7), how responsibility for managing health risks should be distributed between individuals, states, and other agents (Chapter 8), and the nature of health equity (Chapter 9).

If there is both a sufficient evidence base for a policy, and the policy is one that meets the criteria for prioritization within public health, but the government decides not to go forward with the policy, then a reasonable justification for not so doing is required. The most obvious such argument would be that the policy would have a disproportionately adverse impact on some good other than health that the community has reason to care about. There are a variety of interests that could be invoked by a government in this regard, including, on occasion, commercial interests. However, if the right to public health can justify significant infringements of liberty and autonomy, then it should be relatively easy to see how the same kinds of argument apply a fortiori to interests such as commercial interests, which do not invoke fundamental rights.

6.5 Why the Right to Public Health is Compatible with Reductions of Liberty

Many experience an initial stumbling block with the very idea of a rights-based approach to public health, which would allow government interventions to protect health that interfere not only with individuals' liberty, but also their autonomously deliberated decisions. How could interfering in a person's life without their consent (and even in the face of their informed and autonomous dissent) be justifiable in the name of protecting that individual's rights?

I aim to disarm this line of objection by showing that in other circumstances the protection of rights frequently involves the state restricting the liberty of the holders of the right, and sometimes coercing them, all in the cause of protecting the right. So if public health activity involves governments in restricting liberty or coercing individuals (and even if some individuals violently oppose some of the public health measures that the government adopts), that does not by itself show that these interventions could not be justified by a right to public health.

The right to security is usually interpreted in such a way that it allows (or even requires) extensive interferences in the name of protecting the right. Take mandatory airport security measures, which involve systematic interference with the liberty of those who have not committed a crime or other wrong, and are not even suspected of so doing. Under current regulations, if I am going to fly on a plane, I must submit to the scanning of my luggage and to bodily examinations. Many of the activities that are interfered with for the purposes of the right to security look to be self-regarding if anything is (e.g. carrying a small bottle of water).

If it were correct to think that the importance of autonomy precludes protecting rights by interfering in self-regarding actions, then it would be impossible for such interferences to be justified on the grounds of the right to security. The usual rationale for this policy would be that while there is no specific reason to suspect the vast majority of those who are stopped, and only a miniscule percentage of those who are stopped would have been going to commit a crime, these measures are justifiable in the round because they are the only practicable way of detecting and deterring terrorists who otherwise would be able to violate rights with impunity. Most believe also that such security measures are not just rendered permissible, but are actually mandated by the right to security.

The nature of security as a public good entails that it is difficult to protect the security of all without infringing the liberty of some. Even if *I* wish to waive *my* right to security, this does nothing to waive others' rights to security. Hence, if the state is entitled to interfere (in a proportional manner) with the liberty of anyone in order to protect individuals' rights to security, it can interfere in this way with my liberty even if I wish to waive my own right to security. I may not want to have my security protected by being forced to take off my shoes and belt in order to get on an airplane, but it does not follow from this that the government is not entitled to do so in the course of protecting others' rights to security.[17] Indeed, if there are cost-effective and proportionate measures available that could have foreseeably prevented a threat eventuating and these are not taken, then there will be a case that the government has violated the right to security.

Of course, not just any interference can be justified in the name of the right to security. Many measures are disproportionate, and moreover the drive to securitization of the state can have a number of very severely problematic side effects.[18] For our purposes, I wish to take two things away from this brief consideration of

---

[17] Sreenivasan (2012) seems to assume on this basis that apparent rights such as the right to security cannot be waived, and so could not be rights on the will theory of rights. This seems to mistake what needs to be waivable for there to be a will theory right. Compare a case where all the residents in a shared house have a right (that they can individually waive) that loud music not be played after 11 p.m. If I waive my right, then I extinguish the duty that my housemates would otherwise have had in respect of me not to play loud music after 11 p.m. But the other housemates each separately maintain their rights not to be disturbed by loud music. The fact that my waiving of my right does not give a permission for loud music to be played does not show that I did not have a will theory right.

[18] For a helpful discussion of the foundations of security policy, see Loader and Walker (2007).

security: (1) the claim that a right could not support interference with self-regarding actions is a non-sequitur; (2) even if a given individual would prefer not to have their rights protected in a way that they personally find burdensome or annoying, this would not by itself defeat the claim that such measures were required by rights.

If it is true that preventing the sale of sugary beverages in large containers is required, and that citizens' rights will be violated if the government *does not* do this, where would this leave the individual who prefers to have a 'Big Gulp', consuming a massive cup of Coke or Sprite at one sitting? Has her right to non-interference been violated? We can distinguish three kinds of worries here. First, someone might place a particular value on the ability to buy soft drinks in large containers—it may be that for them then, having two half-litre cups of Coke is problematically different from having a single one-litre cup.

The mere fact of objection—even very vocal objection—to a government policy does not by itself mean it would be wrong for the state to enact the policy. What matters is the grounds on which the objection is made. As Chapter 5.7 considered, all real-world policies will be objected to by some citizens, and because of this it would not be feasible to treat a single objection to a policy as determinative. Due to what Rawls (2001: 35–6) describes as the burdens of judgement, we can expect significant levels of disagreement about the nature of the good life within democracies. Given these pervasive differences, policymaking will inevitably have to interfere in ways that some citizens find objectionable. States need, at least implicitly, to take a view on the significance of choices, as not all choices can be preserved equally, and preserving some choices has the implication of shutting down others. So without being able to articulate why the size of cups that drinks are sold in is of sufficient significance to outweigh the health benefits of the policy, then the claim to have been wronged by the policy is unconvincing.

A second interpretation would be as a more general worry about the state overstepping its proper role. But if that is supposed to be the argument, then it seems to amount to little more than the denial (without argument) of the claim that there could be a right to public health. Clearly, *if* there is a right to public health, then protecting health forms a legitimate part of the state's role as, by definition, acting in a way that is proportionate and necessary to protect rights is not wrongful. As we saw in the case of the right to security, activity that infringes liberty and/or overrules individuals settled wills about their own life is justifiable if necessary and proportionate to prevent violation of rights. Therefore, curtailing individuals' liberty, and even acting against individuals' settled wills can be morally legitimate (where it is necessary to prevent rights violations).

A third interpretation would be that the policy is wrong because it is paternalistic. However, as Chapter 5.6 discussed, it is far from clear if a policy counts as paternalistic towards X, if it restricts the liberty of X (along with all other citizens) in a way that is a proportionate response to a right that all have. In any case, the charge of paternalism cannot provide a powerful objection to proportionate action that is required by the protection of rights. Anything that is morally required is either not paternalistic at all, or paternalistic but not morally wrong.

Overall, I have suggested that we see the right to public health as analogous to the right to security. Infringing personal liberty for the sake of public health is relevantly similar to infringing liberty for the sake of security. Just as significant infringements of personal liberty are justifiable for the sake of security, so significant infringements of personal liberty are justifiable for the sake of public health. Interventions to promote or protect *health* that infringe liberty and/or overrule individuals' settled wills are justifiable in the same kind of way as in the case of the right to security. There seem to be two options for the defender of the non-interference assumption (beyond, as we have already considered, simply denying that there is a right to public health).

1. The interference with liberty in cases like airport security is not in fact permissible (despite its ubiquity). Therefore, we should not infer that other equally significant infractions of liberty are permissible in the cause of promoting public health.
2. There is a relevant dissimilarity between the right to security and the right to public health. Even if it is legitimate to interfere with liberty for the sake of security in the airport case, it does not follow that interferences with liberty are justifiable for the sake of public health.

If someone wishes to make objection (1), then the problem goes much wider than health promotion activities. Rights will be systematically violated by all kinds of security policy and surveillance activities. While much public health activity would be difficult to justify, so would much existing government activity that is widely accepted by those who are in general sympathetic to the non-interference assumption. (What is puzzling and interesting is that the case of interference for the purposes of improving health, and interfering for the purposes of security are usually treated very differently; and that some of those who are most vociferous in their denunciation of state interference to promote health are strong supporters of state interference for the sake of security.)

If someone wishes to make objection (2), this would require them to say what the relevant dissimilarity is. One apparent difference might be that the actions stopped by checks such as the right to security would involve moral wrongs, while those prevented by the right to public health would not. However, if this were the argument, then it would beg the question against the existence of the right to public health: if there is a right to public health, then individuals can be wronged by governments failing to fulfil their duty to reduce risks to health.

6.6 Conclusion

I have argued that there is a right to public health, which entails that individuals have an entitlement that their governments systematically remove threats to human health by undertaking health protection and health promotion measures. While any government action under the right to public health needs to be justifiable as proportional, significant infringements of personal liberty can be justified where this is necessary. This argument is at this stage still fairly abstract, and it is dependent on the following chapters to fill out the details. Chapter 7 specifies the right to public health approach by considering how to prioritize between different risk reductions. Part III completes the picture.

7

Which Risks to Health Matter Most?

7.1 Introduction

One of the distinctive advantages of the right to public health account is that it not only allows, but requires decision-makers to give individualistic justifications of public health policy. It provides an orientation for states, by incorporating ideas of respect for individual rights and autonomy within the fundamental ethical principles of public health. Because of this, it greatly reduces the danger that policies that are, on balance, worse for everyone are imposed simply because they will save lives at a population level.

The right to public health requires not a total elimination, but a proportionate reduction, of health-related risks. Chapter 6.4 briefly considered some cases of 'best buys' for public health, which it is difficult for states to justify *not* implementing. Most of the public health systems in high-income countries will already have implemented all, or nearly all, of these best buys. It will not be news to a public health official reading this book that tobacco advertising should be banned, or that safety restrictions such as seatbelts in cars or motorcycle helmets should be enforced. The important question is how states should make decisions about which health risk reductions to prioritize, in cases that go beyond the obvious.

The right to public health account does not by itself straightforwardly determine what should be done in cases that go beyond the obvious best buys. As Chapters 5 and 6 have stressed, whether a particular policy is required by the right to public health depends on a number of contextual factors such as the potential health benefits it would bring, the availability of resources, the degree to which the policy is either supported or resisted by citizens, and how it compares to other alternative potential policies. Working out which policies should be adopted to protect and promote health is not something that can be determined in a top-down manner merely by brandishing the right to public health. It requires detailed, subtle, and contextual work.

In addition, meeting obligations to individuals under the right to public health does not exhaust the state's obligations in respect of health. Population health risks ought also to be reduced in some cases that go beyond what can be demanded by individuals as a matter of rights under the right to public health. Disease eradication policies provide a good example. In real-world cases such as polio, an eradication policy is likely to have a long 'last mile', in which millions or even billions need to continue to be vaccinated for a prolonged period of time to complete the task, despite the fact that there are very few remaining cases of actual infections. Moreover, the time and resources required for compliance with the demands of an eradication programme could instead be devoted to goals that may be more pressing in a particular country. So, as Chapter 10.5 explores further, where disease eradication policies are ethically justified, this will usually be more plausibly done on the basis of the sheer scale of the health benefits to be obtained, than by the claim that anyone's right to public health would have been violated if a disease containment policy had been pursued instead. Thus, while the right to public health sets obligations that the state must meet in public health policy, and in addition public health policies need to be justifiable to individuals affected by them, not every public health policy needs to be justified as required by the right to public health.[1]

This chapter aims to clarify some fundamental questions about which harms and risks to health are most ethically pressing to address. Much of this will be concerned with providing an analysis of the idea of a *claim*, which has become a philosophical term of art in thinking about the fair allocation of scarce resources. Philosophers often take it for granted that if we know what tends to make one person's claim stronger than another in general, such lessons would also hold with respect to their claims to limited resources within a health system. Much of the argument of the chapter involves looking more closely than philosophers usually do at how to measure the strength of individual claims to receive interventions that would reduce health-related risks.

The strength of a claim sounds simple in the abstract, and is usually treated by philosophers as a black box, but it is in fact the source of significant difficulties. The analysis identifies a number of different dimensions and contextual features of the strength of individual claims that someone who wanted to solve these problems in the abstract would need to take a stand on. In some of these cases, it seems hard to find a general solution to the challenge—and any of the proposed mainstream answers will lead to answers that seem paradoxical enough as to be difficult to believe in some circumstances, and so there are question marks about whether these contributions have internal validity.

These paradoxes and inconsistencies are grist to the mill of many technical philosophers, but they are much more worrying for those whose aim is not just to test their intuitions against outlandish thought experiments, but to gain wise guidance about how to make prioritization decisions in real-world cases. The

---

[1] There is the potential for subtle debates—that I note but do not enter into here—about where to draw the line between what the state should do in virtue of its duties to individuals under the right to public health and what it should do under the broader goal of improving population health. One way of drawing the line would be to ask whether a failure to enact a particular policy would wrong individuals, or if it would be merely regrettable. The United Nations' understanding of the human right to health in terms of a requirement on states to progressively realize the highest attainable standard of physical and mental health makes this line particularly difficult to draw (UN Committee on Economic Social and Cultural Rights 2000).

significant and unresolved disagreement about which elements are relevant to the strength of individual claims, and how to reconcile them, entails that in addition the literature as it stands is likely to lack external validity.

In the face of difficulties, we could adopt one of two broad strategies. First, to double down on the assumption that there is a single correct way of measuring the strength of claims to receive health interventions, and to explain the difficulties exposed by the analysis of the chapter as showing that the problem is that this one correct way has not yet been discovered. Second, to draw the conclusion that there is no single correct way of measuring the strength of individual claims—and to think of the idea of a claim as a way of sharpening and articulating certain questions about prioritization, but given the degree of reasonable disagreement about how to measure and weigh claims, not a way of resolving these questions. Drawing on the analysis of performativity and social construction advanced in Chapter 4, the chapter concludes that only the second option is viable. How to think about concrete problems in health policy in the light of these deep-seated differences in opinion is something we will turn to in Part III.

7.2 Prevention, Treatment, and Rescue

When considered in the abstract, nearly everyone agrees that prevention is better than cure. Prevention is also often much more cost-effective (Owen et al. 2012). Nonetheless, few health systems follow through on these priorities in practice. For example, few, if any, spend more on smoking prevention and cessation than they do on cancer care. This fact raises a question: are most health systems in fact mistaken or unethical in their priorities, or is there more to be said for cure in concrete cases than in the abstract?

Larry Gostin (2014: ch. 14) offers a thought experiment, which provides a useful initial orientation. Gostin (inspired by Rawls's 1999 famous veil of ignorance thought experiment) asks us to imagine a constitutional assembly, in which the deliberators must choose the kind of health system they want for their society.[2] The catch is that none know anything about their society or the place they will play in it. They don't know, for example, whether the state is in the Global North or the Global South; their gender, whether they will end up being healthy, or ill or disabled; or whether they will be living in a rural or urban environment. As in Rawls's original thought experiment, the hope is that stripping away knowledge about who we are—and what might be likely to benefit us at the expense of others—provides a powerful way of removing bias.

---

[2] Gostin's thought experiment differs from Rawls's in that Rawls focuses on the principles by which the basic institutions of society will be run, but Gostin focuses only on the health system.

Given this rather unusual decision situation, would you choose to prioritize provision of healthcare, or population-level strategies that will improve public health and the broader social determinants of health? Gostin stipulates that you must choose between two options. Option One would strongly prioritize provision of healthcare:

> You could see a health care professional whenever you wanted to, attend high-quality clinics and hospitals, and gain access to advanced medicines. This scenario would achieve the ideal of universal health coverage but would be highly oriented toward medical care—leaving gaps in population-level public health services and the social determinants of health. (Gostin 2014: 420)

This option, Gostin notes, would 'best serve the interests of individuals already ill and suffering, but it would have limited impact in preventing illness, injury, and early death' (Gostin 2014: 420). Option Two would prioritize preventative strategies at the population level:

> As a result, everyone would live in an environment in which they could turn on the tap and drink clean water; breathe fresh, unpolluted air; live, work, and play in sanitary and hygienic surroundings; be free from infestations of mosquitoes, plague-ridden rats, or other disease vectors; not be exposed to tobacco smoke or other toxins; and not live in fear of avoidable injury or violence. This scenario would make unsparing use of public health measures but would offer no assurance of medical treatment. (Gostin 2014: 420)

Gostin argues that a rational deliberator behind the veil of ignorance would 'unhesitatingly choose to inhabit an environment with healthy living conditions…than gain access to medical care after becoming ill' (Gostin 2014: xv).

This may seem plausible, but perhaps also too easy. As we noted, those setting health priorities around the world in fact very often prioritize treatment over prevention. As Chapter 3 argued, where there is a discrepancy between the results of a thought experiment and what seems ethically appropriate in real-life situations, this could just as easily indicate a case of failure of external validity, as indicate that the thought experiment allows a more perspicuous view of the ethical problem. Perhaps there is more to be said in favour of choosing medical care over prevention than Gostin allows in real-life cases.

Real-world cases where people require urgent medical attention provide a useful counterpoint. When a child falls down a well, or miners get trapped below ground, hardly anyone thinks it a bad thing if a massive rescue operation is mounted, even if the money spent on the rescue could have prevented significantly more deaths if devoted instead to cost-effective preventive measures. Indeed, it is often argued that there is a special moral obligation to save those who are in peril even though more good could be done if we were to deploy our resources more prudently. Albert Jonsen christened this response to such cases the rule of rescue, though he wisely noted that 'The imperative to rescue is, undoubtedly, of great moral significance; but the imperative seems to grow into a compulsion, more instinctive than rational' (Jonsen 1986: 174).

It is far from straightforward to work out what should follow from our response to such rescue cases. Whether a situation is in practice framed as one for which a rescue response is required depends heavily on factors that seem to be ethically arbitrary. Rescue responses are more likely to be triggered by a large percentage of a small group in peril than a smaller percentage of a larger group.[3] They are triggered by identifiable individuals (such as a group of trapped miners) rather than statistical lives (such as the number of miners who could be saved if we were to put in place better safety arrangements for the future). And they are more likely to be triggered by a sudden drop in well-being for the group than by a gradual drop, or by the fact that the group has been in a bad way for a very long time (Jenni and Loewenstein 1997; McKie and Richardson 2003). None of these seem like very cogent ethical reasons.

Overall, the psychological effects of the impulse to rescue depend on framing the situation as an *exceptional* one in which a sudden and unexpected calamity is avertible by swift action. But it is mistaken to view the design of routine public health policy through this lens, as patterns of morbidity and mortality are not generally unexpected at a population level (Cookson et al. 2008). For any given person, a stroke or a road traffic accident may be a sudden and unexpected catastrophe, but epidemiological research allows those running health systems to predict—within fairly narrow confidence intervals—the expected number of strokes or road traffic accidents per year. It is simply a dereliction of a government's duty if it does not collect such data, and plan policies on the basis of the expected number of cases.

Faust and Menzel (2011) distinguish three possible positions on treatment versus prevention, which will help us to bring both Gostin's thought experiment, and questions about ethical obligations to rescue, into better focus. First, prevention and treatment could be (other things being equal) ethically equivalent. On the equivalence view, 'no preference should be given either to preventive or treatment services above and beyond what other allocation criteria such as efficiency, effectiveness, and compensatory justice generate' (Faust and Menzel 2011: 15). Second, other things being equal, treatment could enjoy an ethical priority over prevention, so that decision-makers should prefer treatment 'when the two are equally effective (or cost-effective) in producing health benefits, and even, up to a point, when treatment is *less* effective (or cost-effective)' (Faust and Menzel 2011: 15).

---

[3] This is something that charities have learned to use to their advantage: people will give more if they are asked to help a single child than to help many thousands of people who are suffering.

Adherents of the rule of rescue would seem to need to argue in favour of treatment priority. While rescuing will sometimes be the most cost-effective thing to do, it would only be true to say that there is a specific ethical importance to rescuing if this sometimes required us to prioritize rescues that were less cost-effective than other available interventions. So the ethical cogency of the rule of rescue requires, but does not immediately show us how to justify, treatment priority.[4] Finally, other things being equal, prevention could enjoy an ethical priority over treatment, so that decision-makers should prefer prevention over treatment 'when it as effective (or cost-effective) in preserving health as treatment is in restoring health, and it should have priority even, up to a point, when it is *less* effective (or cost-effective)' (Faust and Menzel 2011: 5).

Although Gostin does not state it explicitly in his thought experiment, it seems intended that the amount of resource that can be deployed is equal between Options One and Two. Thus, the question could be rephrased as: given a fixed and limited budget, how much of that budget should be used to provide access to healthcare, and how much should be used to provide population-level preventive interventions? If so, this would mean that the thought experiment unfortunately does not adequately distinguish between two potential reasons for favouring his Option Two.

First, those contemplating the thought experiment may be responding to the background thought that *population-level measures are more cost-effective*. On this view, prevention would be preferred to treatment because by doing so, the burden of disease will be reduced more cost-effectively. This would be neutral between the equivalence view, and prevention priority. Second, choosers could be expressing prevention priority—namely that prevention is to be preferred to treatment even if it is no more cost-effective. On the latter view, even if decision-makers can create the same total health benefit or the same total reduction of the burden of disease for the same amount of money by treating cases of lung cancer as they could do by preventing lung cancer cases, then they should choose prevention.

You might think that the thought experiment does nonetheless provide a reason to favour prevention priority—perhaps because of the importance of health and of maintaining health. However, this is not the case. If we assume that the cost-effectiveness of treatment and prevention is the same, then the opportunity

---

[4] Sheehan (2007: 359) argues that a more cogent moral justification for the rule of rescue may be available: 'the fact that we are strongly inclined to use large amounts of resources (or otherwise risk great cost to ourselves) in order to save an identifiable individual suggests that we have a *prima facie* agent-relative obligation to those in need of rescue...we stand in a special relationship, perhaps a relationship of circumstance, to those in need of rescue and as such have a *prima facie* obligation to save them.' However, the psychological propensity to rescue favours the existence of this moral duty only if the existence of the moral duty would provide a more convincing explanation of the psychological propensity to rescue than other accounts. Sheehan does not provide an argument for why this should be.

cost of ensuring that one individual has their disease prevented under the prevention approach is that someone else loses the same amount of health because of a lack of availability of healthcare. So the goodness of functioning and the importance of health do not by themselves favour preventive measures over restorative measures of the same cost-effectiveness.[5] In short, Gostin's thought experiment does not seem to give us reasons to go beyond an equivalence view.

One fruitful way to move forward is to question the idea that rescue versus non-rescue or treatment versus prevention provides an adequate conceptualization for the purposes of making fair decisions in public policy. Things may be better characterized by attention to continuity and interconnection. Many chronic conditions, if not adequately controlled and treated, will lead to exacerbations and the need for emergency care; and so it is potentially short-sighted and self-defeating to prioritize 'rescue' over altering the systemic conditions that lead to the requirement for such rescues. Moreover, all healthcare interventions are preventative: they aim either to spot the signs of stress or disease early in order to treat before things become worse; or they are focused on a patient who is already feels ill and needs help to prevent the effects of the illness either lingering or becoming even worse.

Within public health, it is often claimed that there are three levels of prevention: primary prevention aims to reduce the likelihood of disease or injury before it occurs (for example, antismoking campaigns); secondary prevention aims to diagnose disease early in order to allow interventions that will minimize its effects (for example, cancer screening; training of employees to spot the signs of workplace stress); and tertiary prevention aims to minimize the effects of disease or injury that is already severe enough to have made a noticeable difference to the patient's life (for example, rehabilitation programmes after a stroke; antiretroviral drugs after an HIV diagnosis).[6] Thus, the core decisions are not ones of allocating resources to treatment *as opposed to* prevention, but allocating resources between primary, secondary, and tertiary prevention.

---

[5] If you doubt this point, it might be that you imagined deciding about one and the same individual, asking yourself 'would it be better if *this* individual does not get a disease at all or that they get the disease and are cured?' But decisions at a policy level are very rarely decisions about single individuals. The choice would be between:

- Some individuals do not get ill in the first place because health risks have been removed from the environment, but fewer people who do get ill are able to be treated, OR
- More people get ill because certain health risks have not been removed from the environment, but more of those who do get ill are able to be treated and restored to health.

Assuming that the effects on the total burden of disease are the same, then it would be far from clear that the importance of health would favour Option One over Option Two. One other argument made for preferring prevention over treatment is that treatment is usually more unpleasant to receive than preventive measures. However, preventive measures often need to be applied to many individuals in order to prevent one case, and so may on occasion do more harm than good, as Section 7.8 discusses.

[6] Sometimes, a fourth, quaternary prevention is added—where this aims to minimize harms from overmedicalization and from secondary or tertiary prevention (Martins et al. 2018).

7.3 Pairwise Comparison and Aggregation

Although features such as the size of the group in peril, their identifiability, and the suddenness of the drop in well-being are generally not thought to have sufficient intrinsic ethical significance to justify treating rescue cases as different in kind from others, there are four elements of classic rescue cases that have been argued to have significant independent ethical plausibility even in a policy context. These are, first, that the people who require rescue are very badly off—and improving the position of people who are very badly off has been argued to be *much* more important than improving the position of people who are better off. Second, the potential benefit to each individual is very large, and it is argued that it is more important to provide a very large benefit to a small group of people, than a much smaller benefit to a larger group of people. Third, the harm will occur *imminently*—and it has been argued that it is significantly more important to save a life *now*, than to take steps now that will lead to a life being saved in ten or twenty years' time. Fourth, the risk is *concentrated*—and it has been argued that, other things being equal, it is worse if one hundred people are subjected to a one-in-a-hundred risk of death, than ten thousand people being subjected to a one-in-ten-thousand risk of death, despite the fact that the expected result would be exactly one death in both cases.

We will examine the cogency of each of these considerations in more detail, but before doing so, we need to address a more general question about what morality requires when we are dealing with limited resources and how to determine this, which underlies this whole debate. The literature has been dominated by a dialectic between two competing visions. The first vision is that of utilitarianism, or consequentialism more broadly. Utilitarians assume that benefits and burdens, risks and harms are all comparable, and can all be added up in an all-encompassing decision procedure, and that what is required for ethical decision-making is that the decision taken is expected to realize at least as much value (as defined by the ranking procedure) as any available alternative act.[7] Thinking in this way pushes us towards a position where *all* benefits, harms, and risk reductions matter, even if the effects on each individual are tiny. If all harms and benefits can be aggregated, then an arbitrarily large number of tiny benefits will outweigh saving a life.

One initial objection to such a view that I mention only to put it aside is that some harms are so trivial that they should not count at all, ethically speaking. On such a view, only risks or harms that meet a threshold level of severity should

---

[7] The range of ethical positions which allow for aggregation of benefits and harms—as will be considered in Section 7.4—is wide and will include aggregation functions that weight benefits to those who are worse off much more heavily than those who are better off.

count. Such a view has very implausible implications, and so we should not believe it. If each miniscule harm did not count, then a series of miniscule harms would not count either. But many very significant harms (such as those due to climate change) are caused by a large number of separate events, each of which individually has a negligible effect (Glover 1975; Parfit 1984: ch. 3) Similarly, it would be foolish to disregard tiny risks. The risk of any particular bacterial cell mutating in such a way as to become resistant to a new antibiotic may be trillions to one; but given that even in the USA alone it has been estimated that antibiotics are applied to thousands of trillions of cells each year, such mutations will occur as a matter of course (Drlica and Perlin 2011: 76). These cases indicate that it is implausible to think that some harms or some risks are too small to count, ethically speaking. It would be wrong to say that they are of *no account*, though it might be plausible to say that tiny harms or risks can often be set aside as *not relevant* to decisions about much larger harms or risks.

The second competing vision is nonconsequentialism. Nonconsequentialists frequently argue that reasoning which allows many small benefits to outweigh a large harm ignores what Rawls (1999) describes as the *separateness of persons*: in assuming that it is only the overall magnitude of benefits and harms that matters, it fails to take into account the significance of *who* the benefits and harms accrue to. For example, if a sheriff allows an innocent man to be hanged in order to pacify a baying mob, this may conceivably correctly be more conducive to overall utility than other available actions, but it has long been recognized as a paradigmatically unjust act, and for that reason morally impermissible.

In an influential intervention, Thomas Nagel argued that the equal moral status of human beings is better respected by a very different approach from the overall aggregation of burdens and benefits, which he labelled *pairwise comparison*. In pairwise comparison, each person's point of view is considered separately and individually, and compared to all other relevant points of view, and priority is given to fulfilling the strongest individual claim. The idea of a *claim* is a term of art within the literature on pairwise comparison, and we shall unpack some of these debates shortly. In the meantime, it is important to note that claims can be either weaker or stronger. It is treated as axiomatic that if it would be possible to meet only one of two claims, it would be unjust to give preference to meeting a weak claim over a strong claim. Pairwise comparison is claimed to better respond to the separate and equal moral status of human beings than alternative approaches, because the solution it recommends in a case of competing claims will be 'least unacceptable to the person to whom it is the most unacceptable', and thus 'any other alternative will be more unacceptable to someone than this alternative is to anyone' (Nagel 1979: 123).

Scanlon (1998) gives the following much discussed case, which will allow us to clarify both the idea of a claim, and the difference between utilitarian and pairwise comparison approaches:

Suppose that Jones has suffered an accident in the transmitter room of a television station. Electrical equipment has fallen on his arm, and we cannot rescue him without turning off the transmitter for fifteen minutes. A World Cup match is in progress, watched by many people, and it will not be over for an hour. Jones's injury will not get any worse if we wait, but his hand has been mashed and he is receiving extremely painful electrical shocks. Should we rescue him now or wait until the match is over?   (Scanlon 1998: 235)

From the way that the case is set up, it is obvious that we are supposed to draw the inference that Jones has a much stronger claim to be relieved from receiving the electric shocks than anyone else has to watch the match. Both pairwise comparison approaches and approaches that allow aggregation will agree that it would be ethically much more urgent to relieve Jones of his pain than to allow a single individual to watch the match. For those such as utilitarians who allow aggregation, the numbers do matter, and if there are enough people who would have to forgo some watching of the match, then it would be better to tell Jones to tough it out. On the pairwise comparison account, the numbers are not relevant: regardless of how many viewers there are, we should not make Jones wait.

However, if aggregation is *never* allowed, then pairwise comparison will have implications that are just as counterintuitive as those of always aggregating. The most radical advocates of the non-aggregative picture follow Taurek (1977) in claiming that the numbers of people who have claims of a particular strength are never ethically decisive. On Taurek's view, if we face a choice between saving either a smaller group, or a distinct larger group, from a given harm, then there is no ethical requirement to save the greater number. Many find it difficult to accept the implication that even if all the members of both groups are innocent and face imminent death, and the second group is arbitrarily larger than the first, then there is no ethical requirement to save the larger number.

Many nonconsequentialists disagree with Taurek and argue that numbers do matter where the strength of claims are equal. But the pairwise comparison approach in its pure form has other implications that are just as difficult to believe amount to genuine ethical insights. The virtue of the response to Jones and the TV transmitter case was supposed to be that where one claim is stronger than another, the numbers do not matter: the right thing to do does not depend on the numbers who are watching the match. But this threatens to have the disquieting implication that it could be ethically required to leave an arbitrarily large number of slightly weaker claims unmet in order to allow one somewhat stronger claim to be met. Depending on the precise way in claims are measured, this may have absurd, or morally repellent consequences. If, as we discuss in more detail later, someone who faces certain death has a stronger claim that someone who faces a 50 per cent chance of death, it would require prioritizing saving a single individual who would otherwise certainly die over saving millions who would have a 50 per cent chance of death, even if the death would be equally as unpleasant for the individual where they occur.

Given these difficulties, one popular strategy, developed in detail in Voorhoeve (2014) is to combine pairwise comparison with aggregation. On this approach, weaker claims should be aggregated and allowed to outweigh stronger claims through sheer force of numbers only where the weaker claims are strong enough in comparison to the stronger claim to be *relevant* to it. Thus, the claim to treatment for a mild rash is not relevant to the claim to treatment for cancer, so there is no number of mild rashes that could outweigh one life-saving cancer treatment. However, perhaps chronic lower back pain will be bad enough that the claim to treatment for it is relevant to the claim to be treated for cancer, and so there will be a number of back pain treatments that would outweigh one life-saving cancer treatment.

This sounds plausible, but there are a number of deep theoretical and practical worries with the proposal. The main theoretical worry is that the idea of relevance seems ad hoc, and potentially contradictory, and does not seem to survive well when put under pressure in complex cases. First of all, there are obvious problems of transitivity. Suppose that (1) claims of strength A are weaker than claims of strength B, but still relevant to them; (2) claims of strength B are weaker than claims of strength C, but still relevant to them; and (3) claims of strength A are not relevant to claims of strength C.

According to the aggregate relevant claims view, there will be a number of claims of strength A that will outweigh a single claim of strength B. As claims of strength B are relevant to claims of strength C, there will be a number of claims of strength B that will outweigh a single claim of strength C. However, there is no number of claims of strength A that will outweigh a single claim of strength C. This looks odd, and it seems we will end up with path-dependent results: if in a situation there are many more claims of strength A than strength B, and many more claims of strength B than strength C, then (assuming the numbers are sufficient) we will face the following intransitivity: we should choose A over B, B over C, but C over A.[8]

Moreover, the account also is problematic for other reasons: it violates the *Independence of Irrelevant Alternatives*. It turns out that adding additional options that would not in any case be chosen can make a difference about which of two options should be chosen (Halstead 2016). In an ingenious paper, Patrick Tomlin argues that things get worse from here, as the Aggregate Relevant Claims view gives unstable results in cases where we are deciding between multiple claims of different strengths. Tomlin points out that there is an ambiguity in the idea of relevance: in order to be relevant, does a claim need to be strong enough relative

---

[8] Voorhoeve is aware of this implication, but does not think it devastating to the theory. For further discussion, see Privitera (2018).

to the strongest claim it is competing with (*anchor by competition*), or strong enough relative to the strongest claim in the whole situation (*anchor by strength*)? Either way, the Aggregate Relevant Claim account seems to imply results that seem very hard to believe. In some circumstances, it turns out that adding the same number of weak claims of the same strength to both sides of a tied competition between competing claims will shift the balance; in other circumstances, adding additional claims only to one side of an initially tied competition will tip the scales in favour of the *opposing* side (Tomlin 2017).

Overall, it is safe to say that the philosophical literature on aggregation and competing claims fails, despite its sophistication, to provide compelling ethical guidance even in the highly simplified thought experiments that are traded back and forth. Both never aggregating and always aggregating lead to problematic implications; but the attempt to marry the two by aggregating only relevant claims struggles to avoid an incoherent syncretism. In other words, to use the distinction introduced in Chapter 3, there is good reason to doubt that these thought experiments and theories have internal validity, and their claims to external validity must also be shaky. It is perhaps no surprise that Norman Daniels argued that, in the light of the apparently insoluble paradoxes and disagreements, it would be better to stop expecting to solve such problems with pure philosophy, and instead open up the ethics of resource allocation in healthcare to a procedural approach that emphasizes accountability for the reasonableness of decisions taken (Daniels 1994, 2007).

Daniels is correct to think that these debates are unlikely to be resolved any time soon; but it may be too quick to jump to a procedural approach. If the problem is that philosophers and ethicists who have been thinking earnestly about these problems for many years have failed to come up with an approach that avoids both very unintuitive implications and is also internally consistent, then it is initially hard to see how it would be better to turn decision-making over to those who are using less sophisticated ethical theories and are no doubt less aware of the limitations of the theoretical commitments that they have (Ashcroft 2008). The failure of professional philosophers to reach a consensus does not mean that anything goes, or that professional philosophers should depart the field in favour of those without philosophical training. As Chapter 1.1 argued, neither politicians, nor citizens, nor corporations will be free from potential conflicts of interest in their attempts to articulate how competing values should be reconciled in the public interest.

Rather than giving up on philosophical approaches, the way forward is to be clearer about the nature of the contribution philosophy can make. Clarifying the implications and the limitations of ethical principles and ethical theories is vital, but we should not assume that as a result of such arguments and clarifications that one single correct view will be left standing. The next sections of this chapter examine four dimensions that are widely thought to be relevant to the strength of claims: priority to the worst off, capacity to benefit, time, and risk. Health systems, informed by this analysis, will need to take a view informed by democratic deliberation of where to position their priorities within the space of defensible and consistent views. As we shall see, there are significant difficulties in coming to a definitive interpretation of any of these important ideas; and so the final section of the chapter and much of Part III look in more detail at the nature of the choices that healthcare systems need to make in specifying what is relevant to the strength of a claim, and how to measure this.

7.4 Priority to the Worst Off

Philosophers typically assume that plausible ethical theories must give a (possibly limited) *priority to the worst off*. Other things being equal, the worse off someone is, the stronger their claim to help; and the worse off someone would be made by an action or intervention, the stronger the complaint they have against it. Acting in accordance with a principle of priority to the worst off requires specifying what we mean by 'worst off', as well as being prepared to advance an account of how to measure this.

In the context of public health policy, one fundamental question is whether 'worst off' should be interpreted as worst off *now*, worst off over a complete life, or in a different way.[9] A simple example suffices to sketch the problem. Suppose that Barry has had a wretchedly painful life of physical pain, mental anguish, and deprivation, but is at the moment calm and content. Susan has led a long and charmed life of exceptional fulfilment and accomplishment, but has just been hit by a car and is in great pain. She is receiving urgent medical attention, which is expected in time to allow her to make an almost full recovery. Right at this moment, Susan is significantly worse off than Barry; but from the perspective of their whole lifetimes, Barry is significantly worse off than Susan. Indeed, she may have enjoyed so much more lifetime well-being than Barry that there may be nothing that we could do to improve Barry's life that would bring him close to the overall level of well-being that Susan has enjoyed.

Interpreting 'worst off' either solely as worst off now, or solely in terms of worst off over a whole life would have problematic implications. Focusing only on whole lifetimes would make health systems insensitive to present needs in a range of obvious cases. It would seem inhuman—and frankly bizarre—if a health system were, for example, to prioritize treatment of a minor ailment for Barry over Susan's lifesaving treatment, on the grounds of priority to the worst off over a

---

[9] This is not the only relevant fundamental question: another is whether for the purposes of public health policy, worse off should be interpreted solely in health terms, or all things considered (Pratt and Hyder 2016). I discuss this question further in Chapter 9.2.

whole life. But equally, a health system that only focused on those who were worst off *now* would be similarly indefensible. Many diseases progress predictably if left untreated. It is counterproductive, and unethical, to construct a system that in effect tells patients to come back when they are sicker. Moreover, there are reasons to think that a system that continually prioritizes those who are worst off *now* will end up worse for everyone than a system that takes a longer term perspective (Gustafsson 2015).

It is thus a mistake to think that there is a single point in time (whether it is now, or the whole life of an individual) from which all decisions about who is worst off should be made. A possible response would be to find some way of combining a measure of worst off now with that of worst off over a lifetime. However, even if a dual perspective were adopted, which somehow merged the perspective of now with that of a complete life in a satisfactory way, this would do little to resolve a more fundamental problem. The chief purpose of constructing public policy is to shape systemic processes that affect citizens' well-being over the medium to long term, rather than to determine what to do at one isolated moment. Static approaches, which look at well-being in time as if it consisted of snapshots, are much less useful for public policy than dynamic approaches which focus on lives as unfolding processes, and on systemic interconnections as they affect individuals through time.

In countless other cases in public health, how well off individuals are relative to one another will change over time, sometimes in fairly predictable ways. To take a very simple example, suppose we can give a health-related good to either to A, B, or C. A is currently worst off; B is better off than A, but worse off than C. C is currently best off. The good will restore one individual to full health. However, while A's condition remains constant over time, that of B and C progressively worsen. C's condition will worsen more quickly than that of B, and if no intervention is made, the severity of A's condition will be first overtaken by B's, and then later severity of both A's and B's conditions will be overtaken by C's. Suppose we are aware of all of these facts, and that all other factors are equal, and we can only give the good to one of them.

If we approach priority to the worst off as by assessing a snapshot at a particular time, it would imply that A, who is worst off *now*, has a stronger claim than B or C who will foreseeably become worse off than A is now if they do not get the treatment. But if we approach questions of well-being and of priority to the worst off in dynamic and structural terms, we will focus on processes, and are likely to take a more holistic view that understands current well-being in the light of a trajectory over the arc of a developing life. This need not translate simply into the idea that therefore C should be treated as the worst off, but rather in a call to rethink the types of intervention adopted over the medium to long term. Chapters 9 and 10 examine some of these questions in greater depth in the light of concrete policy decisions about health-related inequalities and communicable disease, arguing that it is crucial to design policy in a way that is suitable to control risks that are mediated by complex systems and whose nature is partly performative—and that static models are poorly equipped to do this.

7.5 Capacity to Benefit and Opportunity Costs

Quite separate from the question of priority to the worst off, many also believe that how much individuals would benefit from an intervention makes an ethical difference that is relevant to priority-setting. Call this feature capacity to benefit. Capacity to benefit can either be absolute, or relative to financial resources. Where capacity to benefit is understood in absolute terms, the point is that—whatever we do—one person will receive more benefit from a particular intervention than another. As Daniels puts it:

> Suppose that Alice and Betty are the same age, have waited on queue the same length of time, and will each live only one week without a transplant. With the transplant, however, Alice is expected to live two years and Betty twenty. Who should get the transplant?   (Daniels 1994: 27)

Where some people are more efficient convertors of healthcare resources into health benefits than others, it may be unfair if a decision-maker attempts only to maximize the amount of health benefit they produce. Daniels describes this as the fair chances/best outcomes problem. In cases of this kind, there is disagreement in the philosophical literature about whether a morally motivated stranger should flip a coin between giving the transplant to Alice or Betty; or whether the fact that Betty will benefit much more entails that Betty has a significantly stronger claim, and thus the transplant should be given straight to Betty; or whether a weighted lottery should be used, that would give both a chance of attaining the transplant in line with the strength of their claims.[10]

There is a difference in relative capacity to benefit if a similar size of benefit can be provided to two different people, but the opportunity costs of providing it to one are much higher than those of providing it to the other. Expensive treatments for rare diseases provide a good example. The underlying research to bring a drug to market for a rare disease is often just as expensive as that for a more common disease. But given the rarity, the pharmaceutical company will be expected to sell only a fraction of the total number of units of the drug for the rare condition. The

---

[10] A similar worry arises if we approach the problem from a utilitarian perspective: arguably, simply maximizing the outcomes fails to treat Alice with sufficient respect. As Nozick (1974: 41) put it, 'Utilitarian theory is embarrassed by the possibility of utility monsters who get enormously greater sums of utility from any sacrifice of others than these others lose. For, unacceptably, the theory seems to require that we all be sacrificed in the monster's maw, in order to increase total utility.'

research and development costs (plus a profit) must be recouped from this small volume of sales, leading to much higher prices.[11] McCabe provides a useful stylized case:

> Consider two groups of people who have similar diseases (J and K). J is a rare disease (1 per 10 000) and K a more common disease (1 per 1000). Imagine these people have the same personal characteristics, the same prognosis without treatment, and the same capacity to benefit from the treatments...[T]he cost of the orphan drug for J is higher than the treatment for K. Suppose the cost of treating one case of J is £1000, the cost of treating one case of K is £100, and the budget is £1000. Then the real choice posed by orphan status is between treating 1 person with J or 10 people with K.   (McCabe 2005: 1018)

If looked at purely through the lens of the thought experiment, we might conclude with McCabe that 'The idea that decisions should be made based on valuing health outcome more highly for no other reason than rarity of the condition seems unsustainable and incompatible with other equity principles and theories of justice' (McCabe 2005: 1018). The fact that a particular disease is rare (or that it is common) does not appear to be per se ethically relevant.

However, there may be other reasons that are not easily captured within the net of the competing claims account for being willing to pay more for drugs for rare diseases. Pharmaceutical companies will never do the research and development necessary to develop drugs for a rare disease unless they have some assurance that they will be able to be reimbursed. The price of drugs usually falls significantly once the patent expires, and so 'Paying high prices today for rare disease drugs enables future low prices on the same drug' (Hollis 2006: 160).[12] Moreover, having a rare condition that happens for this reason to be expensive to treat is a paradigm case of bad luck, and it has been argued that 'society has a moral obligation not to abandon individuals who have had the bad luck to be affected by a serious but rare condition for which no treatment exists' (Gericke et al. 2005: 165).[13]

---

[11] I examine ethical issues in hyper-expensive treatments in much more detail in Wilson and Hunter (2011).

[12] In practice, markets for generic medicines have often failed to come close to the envisaged perfectly efficient market of economists due to weak competition and limited numbers of manufacturers for many generic drugs. Where a market for a drug remains a de facto monopoly or duopoly, manufacturers have significant scope to greatly increase prices notwithstanding the lack of patent protection. For more on this, see Dave et al. (2017) and Ferrario et al. (2020).

[13] The idea of 'bad luck' is slippery in this context. If it is bad luck to be denied care because you have a rare disease, why could not the person with the common disease also claim to be a victim of bad luck? Arguably, through no fault of their own, they have a condition which is treated by the health system as less worth investing in, simply because it is common (Charlton et al. 2017). I discuss the relationship between luck and justice in more depth in Chapter 8.5.

7.6 Time and Claims

Suppose we can either save ten lives now, or take an action now that will save ten lives in 30 years' time. Which should we choose? Nearly everyone would choose to save the lives now. How about if we would be expected to save more lives (say 15) in 30 years' time? At this point things get more complex. It is important to disentangle the main ethical reasons for this, which include epistemic reasons, commodity discounting, and pure time preference.

One important reason for favouring saving the lives now would be epistemic. It is less certain that we will be able to save the 15 people in 30 years' time. Perhaps a technology will have been invented by then that would allow the lives to be saved even if we do not act now. Less optimistically, perhaps there will have been a revolution or an ecological disaster that will prevent people from doing what is required in 30 years' time; or perhaps those alive then will simply forget to do what is required. Because of this uncertainty, it is often argued that it is appropriate to discount future benefits.

As we saw in Chapter 4's discussion of complex systems approaches, and as will be a key theme of Part III, the requirement to take account of dynamic change and uncertainty is the usual case rather than a rare exception in public policy. However, competing claims approaches are poorly calibrated and underspecified for such cases. In a fast-moving scenario such as an unfolding natural disaster, uncertainty and shifts in estimates of strengths of individual claims will become a factor over hours rather than years. Suppose that a severe storm is due to hit a country within hours—and the resultant flooding, and damage to buildings, and threats to life will play out in the hours and days after the storm hits. How these risks will play out will depend both on how the unfolding disaster goes, and also on the human response to it. Let us further suppose that before the storm hits, policymakers can assign some probabilities to how the event will affect different interested parties, and how bad it will be for them, and how these individuals would benefit from different interventions. So, before the storm unfolds they are in a position to make some *ex ante* adjudications of the strength of different individuals' claims and of who is worst off. As the event unfolds, so will their understanding of the risks and the possibilities. How does the strength of individuals' claims alter as the scenario unfolds? Is the strength of claims set *ex ante* and then remains fixed during the event, or do they shift on a day-to-day or potentially even moment-to-moment basis as the scenario unfolds?

Although philosophers do not often discuss this kind of case, it would be odd for those who believe in the competing claims account *not* to think that the strength of individuals' claims to assistance alters with the severity of the threats to well-being and the possibilities for action to remediate these threats. Shifting plans and priorities in the light of new evidence is a basic requirement of effective disaster response. For example, perhaps in the evacuation plan for part of a city, it was reasonably believed that citizens in one part were at relatively low risk because they could escape via the bridge to safety, but this bridge is then swept away in the storm surge. In this case it may turn out that a large number of people who were previously thought to be at fairly low risk of harm are in fact at high risk of catastrophic harm—unless something is done about this. It would be decidedly odd to continue with one's initial risk management plan in such a scenario, and insist that these individuals should be treated as having a relatively low claim on assistance.[14] As I explore further in Chapter 10.3, where the strength of the claim that an individual will have in the future depends on choices that are made by others in the interim, the strength of this claim is performative rather than being fixed objectively.

It is standard in economics also to apply discount rates to commodities for non-epistemic reasons. The price of most commodities falls over time relative to the return we could get on an investment at a bank, so buying a commodity today has the opportunity cost of the greater amount of the same goods we could buy in the future. The price of commodities falls over time relative to the return on a bank investment both because we become more efficient at manufacturing commodities over time, and because natural resources such as forests grow naturally if we wait before harvesting them. Broome dubs this the fertility of technology: as he puts it, 'present commodities can be converted into a greater quantity of future commodities, if we choose' (Broome 1994: 139). Given these facts, it makes sense to employ a discount rate for commodities because we will be able to buy more of those same commodities for our money in the future.

The economists' discounting model assumes that the increased number of commodities we would be able to buy in the future have the same value for well-being as the smaller bundle we can buy now. However, not all commodities vary in this way relative to well-being: some commodities contribute a constant amount to well-being whenever they occur. Broome's example is saving a life:

> Lifesaving in the future will make the same contribution to well-being as lifesaving in the present. Certainly, future lifesaving is cheaper than present lifesaving, but this is not a reason for valuing it less. The market prices of commodities only have a role in valuations because they measure the relative values of commodities to people. In equilibrium, they will do so... But if lifesaving produces constant well-being and yet is cheaper in the future, we evidently do not have an equilibrium.   (Broome 1994: 150)

---

[14] Obviously, when dealing with a fast evolving situation there will be costs involved in redirecting resources, and it will often be sensible to commit once a decision has been made. For example, in fighting a fire, if already committed to saving a small number of people in one part of the building, it may not be sensible to pull out of this in order to save slightly more people in another part of the building. Where a situation will deteriorate quickly unless something is done about it, and resources to deal with it are limited, rescuers could get into a self-defeating spiral if they continually put aside their current rescues to attend to even more urgent cases.

So, even though we may well be able to create health benefits more cheaply in the future, there is no reason to think that the health benefits thus created should be subject to commodity discounting.

A third reason for discounting is pure discounting: discounting the value of benefits and harms in the future solely for the reason that they are in the future. Most philosophers have followed Ramsey's lead in thinking that pure discounting is 'ethically indefensible and arises merely from the weakness of the imagination' (Ramsey 1928: 543).[15] On this view, the time at which one becomes morally needy ought not to make a difference to the strength of one's moral claim. The reason for thinking this is simple: there seems to be no reason to think that the mere fact that suffering or death is proximal in time provides a reason to prioritize it, any more than there is a reason to think that suffering or death is proximal in space does (Parfit 1984: 356–7). At the point when each person needs rescuing, he or she is equally as needy as any of the others when he or she requires rescue. It is just that different people reach the point at which they need to be rescued at different times.

Overall, while there are good reasons for factoring in epistemic uncertainty into decisions about priority setting, it is much less plausible to think that commodity discounting or pure discounting should be applied to cases of public health.

7.7 Risk and Claims

Risk plays havoc with the idea of pairwise comparison—something that has only fairly recently begun to be explored seriously in the literature. It is so difficult to spell out an approach to risk based on pairwise comparison that is both consistent and gives sensible answers in a range of common cases, that some have openly begun to explore whether the whole pairwise comparison and competing claims project should be abandoned (Fried 2012a, 2019). By contrast, aggregationist approaches have a much more coherent approach to risk: they are already committed to the idea that harms can be summed and compared, and so there is no great paradox involved in thinking, for example, that a 50 per cent chance of the loss of a leg is half as bad as the loss of a leg.

We can distinguish three ways in which being the subject of a risk imposition can be harmful. First, and most obviously, risk is bad because of the eventuation of the harm that was risked. If a risk of death ripens into actual death, then that is a bad thing for the person who dies. Second, awareness that you are subject to risk creates anxiety and undermines a sense of security. Without a sense of security it is difficult to make plans for the future (Wolff and De-Shalit 2007).

---

[15] Arguments for pure discounting have mostly come from economists. For an overview of this debate, see Broome (1994) and Ponthière (2003).

Third, it is sometimes argued that 'pure risk', in and of itself, is a harm. A risk is pure if the threatened harms (a) do not eventuate, and (b) the fact of risk imposition does not enter into the awareness of the person on whom the risk is imposed. A classic example of pure risk would be playing Russian roulette on someone while they are asleep. Suppose that, due to luck, the chamber is empty and the person is unharmed. They also never discover that they have been subject to this 'game'; and it has no other effects. The rest of their life passes off exactly as it would have been had no one ever played Russian roulette on them that night.

There is a small literature on whether the person subjected to pure risk has been harmed. It is not clear to me that the question is a useful one to ask.[16] Cases of pure risk tend to be described in a way that stipulates that they have no effect—other than potentially the mysterious and unexplained harm. So, by the nature of pure risk, it does not seem possible to point to anything about the person's life that is worse other than the mere fact of the risk imposition. So the question seems to be precisely the kind of metaphysical question of which William James and other pragmatists counsel us to ask 'What difference would it practically make to any one if this notion rather than that notion were true?' (James 1907: ch. 2).[17]

Most people judge that if both the severity of the harm (for example, death) and the total expected reduction in harm are held constant, it is more morally important to reduce the risk to a smaller number of people by a larger amount, than to reduce the risk to a larger number of people by a smaller amount. However, it is hard to spell out this ethical claim in a principled way that is consistent and gives sensible answers in a range of common cases.

Suppose a health system has to choose between two options: (1) Save one individual who would otherwise certainly die, or (2) perform a public health intervention that would reduce the risk of death of each of one thousand people by 1 per cent. The first option will save one life, while the second is expected to save ten. But in the first option there is a very significant benefit to a particular individual, whereas in the second case it would appear that the benefit to each individual is much smaller. But for those who *are* amongst the ten who remain

---

[16] This behaviour is certainly morally disreputable: it shows a callous disregard for life, and on that basis is morally to be criticized. It is important to note that the moral criticizability of the act does not depend on being able to show that the person on whom Russian roulette was played was harmed. It is often accepted in other cases that there can be wronging without harming (for example, undetected violations of privacy). So the claim that the person who imposes a pure risk by playing Russian roulette wrongs their victim does not require that they harm them.

[17] Pure risk is in any case not particularly worrisome from a policy perspective, because numbers are large enough that more common risks will eventuate somewhere, and so avoiding actual harms provides an adequate reason for reducing these. The awareness of the likelihood of harm creates duties to take precautions, and individuals can be wronged by a failure to take the expected precautions, even without needing to rely on a concept of pure risk.

alive in the second case only as a result of the intervention, the actual benefit to each of them is presumably very large.

How should we compare and count risks of the same severity but different likelihoods? If we assume that the strength of a claim is determined *ex ante*, by discounting the likelihood of its occurrence, then on a non-aggregative model of pairwise comparison, it will follow that regardless of how many people face a 50 per cent risk of death, it will be more morally urgent to save a single identified person (Smith) from certain death. So, if there is a choice between one million people each being subjected to a 50 per cent chance of death, or one person being subjected to certain death, we should choose to save the one individual—even if the deaths in question would be equally as bad for those to whom they happen.[18]

The situation is importantly different from the TV transmitter case discussed in Section 7.3. That case traded on the fact that each of the viewers would need to lose only something trivial to allow Jones to be freed. In this case, we are aware that many, many individuals will lose exactly the same thing that Smith does (though we do not yet know which individuals these will be). Some will no doubt think that it is ethically required to save Smith in these circumstances, but I find that rather hard to believe.

Shifting to a partially aggregative model does not solve this underlying problem. Presumably a 50 per cent chance of death would be relevant to certain death; but if strength of claim reduces with likelihood then there will come a time when a risk of a particular harm is sufficiently unlikely that it will not be relevant to the certainty of the same harm. Under the theory, if a claim is so weak as not to be relevant, then unlimited numbers of that claim will still be outweighed by a single weightier claim; so it remains the case that the theory is happy to allow unlimited numbers of innocent people to lose exactly the same thing as Smith does in order to save Smith (Frick 2015).

If, alternatively, we assume that the strength of a complaint is determined *ex post* on the basis of what actually happens—then this has, if anything, even more implausible implications. Scanlon argues that 'The grounds for rejecting a principle are based simply on the burdens it involves, for those who experience them, without discounting them by the probability that there will be anyone who actually does so' (Scanlon 1998: 208). Similarly, Reibetanz argues that

> As long as we know that acceptance of a principle will affect *someone* in a certain way, we should assign that person a complaint that is based upon the full

---

[18] A similar result would obtain if, rather than the level of *ex ante* risk that is used to ground the strength of a claim, the size of the risk reduction is held to be relevant (or if both are taken to be relevant). So long as likelihood is used as a way of modifying strength of claims, pairwise comparison approaches will lead to the result that it would be morally required to allow unlimited numbers to die from a threat that has a lower likelihood of killing each individual to save a single individual from certain death.

magnitude of the harm or benefit, even if we cannot identify the person in advance. It is only if we do not know whether acceptance of a principle will affect anyone in a certain way that we should allocate each individual a complaint based upon his expected harms and benefits under that principle.

(Reibetanz 1998: 304)

Given that few, if any, activities are entirely risk free, this seems to have the implication that many everyday activities such as driving or air travel are morally impermissible. We are aware that some people will be killed by them; and whoever will be killed by them will have a complaint against the activity that is far more serious than the complaint at not being able to drive or travel by air. If, like Scanlon, we do not think that claims should be aggregated, then

the number of persons who would have to forgo air travel does not strengthen the complaint over a ban on air travel. This means that the complaint at the exceedingly remote risk of death outweighs the complaint over a ban on air travel. A principle imposing such a ban therefore cannot be reasonably rejected.

(Ashford 2003: 299)

Such a view is widely believed to be too demanding.[19]

7.8 The Prevention Paradox

The debates between *ex ante* and *ex post* approaches have important implications for the ethics of public health (John 2014; Thompson 2018). Most diseases follow a bell-shaped normal distribution in the relationship between risk factors and number of cases. For example, while obesity is a risk factor for type 2 diabetes, and the more obese someone is, the higher the relative risk they will be at, the majority of the cases of type 2 diabetes do not occur in the morbidly obese. Similarly, the majority of those who suffer mortality or morbidity from a disease such as a stroke or heart attack are not in high-risk groups. This is because there are many fewer people in the tails of the distribution.

---

[19] In fact, as Frick (2015) points out, this may be an oversimplification. In cases where there is the possibility of death on both sides of an option, then *ex post* contractualism can actually be fairly conducive to public health thinking: suppose, as in the example at the beginning of this section, we must choose between avoiding certain death for one, and ten statistical deaths. The complaint at not enacting the policy to reduce the statistical deaths will be a full strength one, as there will be someone who will die as a result of it. If it is permitted for the numbers to matter where claims are of the same strength, then it would not just be permitted, but in fact a duty to enact the preventative policy. However, if interpreted in this way, then in many cases, *ex post* contractualism would not be distinctively different from approaches that allow unconstrained aggregation.

One important implication of the normal distribution is that it will often be possible to save more lives more cheaply by reducing the risk slightly for everyone, than by specifically targeting interventions at those who are 'high risk'. Geoffrey Rose argued that because of this, public health should take a population-based approach—in which health ministries try to move the curve as a whole to the left (Rose and McCormick 2001). In adopting a populating strategy, there will still be a right-hand side of the curve, and those in the right-hand side of the curve will still be at higher risk relative to the rest of the population. But those in the right-hand tail of the curve, like the rest of the population, will now be at a lower absolute risk.

One well-known challenge for such an approach is that even though some definitely will benefit from the intervention, we may never be able to identify anyone who did, and it may not be clear that the small risk reduction is 'worth it' for each of the individuals whose lives are altered. In a classic article, Rose explains some of the statistical implications of the long-running Framingham Study as follows:

> If we supposed that throughout their adult life, up to the age of 55 Framingham men were to modify their diet in such a way as to reduce their cholesterol levels by 10%, then among men of average coronary risk about one in 50 could expect that through this preventive precaution he would avoid a heart attack (if change in a risk factor leads to commensurate reduction in risk): 49 out of 50 would eat differently every day for 40 years and perhaps get nothing from it.
>
> (Rose 1981: 1850)

Would it be in your interest to change your diet (or undergo another regular intervention that will slightly reduce the risk something very bad will happen to you, such as taking statins or undergoing breast cancer screening), given the inconvenience? And if you are not sure that it is worth it from a self-interested perspective, is it a good idea to promote such interventions at a policy level?

Rose describes this as the prevention paradox: 'a measure that brings large benefits to the community offers little to each participating individual' (Rose 1981: 1850). The initial temptation in prevention cases is to think about them *ex post*, and rely on the fact that we can identify specific individuals who have been benefited (for example, a particular woman whose breast cancer was diagnosed as a result of screening). This makes the benefits look like those obtained in the rule of rescue type cases. Just as saving one identifiable individual might be thought 'worth it' even given the significant opportunity costs in terms of care or prevention that could have been provided to others, so if screening saves an identifiable individual it is worth all the inconvenience to the others to ensure this benefit.

There are two problems with this analogy. First, the number needed to screen in order to save one life is in the thousands, so saving this one identifiable individual simultaneously means that there are thousands of women who are not in fact benefited and may be made worse off by the screening as a result of false positives (misdiagnosis) or cases in which what is detected is genuinely incipient disease, but where this would not have gone on to cause a clinical problem (overdiagnosis). It is not obvious that it is in the *ex ante* best interests of individuals to take part in the screening (Keen 2010; Marmot et al. 2013), and just as in rule of rescue cases, *ex post* thinking in prevention cases may end up arguably unjustifiably neglecting those who are not identified (Kelleher 2013).

Second, overdiagnosis can occur for a variety of reasons including that the disease would self-resolve, that it would have remained confined to a particular area (e.g. ductal carcinoma *in situ*), or that it is sufficiently slow progressing that the person would have died from another cause before it becomes problematic (Brodersen et al. 2014). In many cases the underlying processes of disease progression are stochastic, and it may be indeterminate whether a given very small cancer would have remained insignificant or if it would have become virulent if left untreated (Wu 2020). It is thus difficult if not impossible even to determine whether overdiagnosis has occurred at an individual level: the extent of overdiagnosis can only really be explained and quantified at a population level (Hofmann 2018). One important implication is that is not possible to identify individuals whose lives *have* been saved as a result of screening. Those whose early stage cancer is detected by screening and are successfully treated may feel that the screening has saved their lives, but no one is in a position to say this, as their treatment could in fact have been overtreatment. So, on closer analysis, even the common assumption that screening *does* benefit identified individuals is contestable.

The alternative is to consider the decision *ex ante*—for example, we can ask whether it would be rational or reasonable for someone who knew the statistics on breast cancer screening, but didn't know if they were going to be one of the lives that was saved, or one of the people who would have an unnecessary biopsy, a false positive, or would simply be given the all clear, to attend the screening.

In some cases it will be *ex ante* worse from the perspective of a representative person to have to undergo the prevention than to be subjected to the risk that the prevention aims to avoid. This would be ethically problematic of a consequentialist perspective that holds a simple aggregative view (in so far as the overall consequences are worse). As Rose put it, 'If a preventive measure exposes many people to a small risk, then the harm it does may readily... outweigh the benefits, since these are received by relatively few' (Rose 1981: 1850).[20] It is also ethically problematic from a deontological perspective, in so far as it imposes a risk on a large number of people that is expected to be worse for them than what it is preventing.

---

[20] More sophisticated consequentialist views may give weight to features such as outcome equality, and thus may recommend the choice of a distribution of risks that is slightly worse for a majority if it improves the prospects of the worst off. For further discussion of these points, and more, see Voorhoeve and Fleurbaey (2012) and Fleurbaey and Voorhoeve (2013).

An interesting problem is posed by cases where there is a choice between a 'high-risk' and a 'low-risk' strategy—where both are believed to be *ex ante* in the interests of those to whom they will be applied (John 2014). If we think that health systems should adopt a policy that targets 'high-risk' patients rather than going for a Rose-like population strategy, then we need to clarify why this is, especially as it will often be the case that the population strategy will be more cost-effective.

One reason would be that those who are at high risk are worse off, and we should prioritize helping the worst off. For example, Norman Daniels argues that equity requires us not only to consider the total number of lives saved, but also how badly off each individual will be if not saved. Where individuals are exposed to a higher concentration of *ex ante* risk, they have a stronger moral claim to be saved. Such an approach could also be used to justify a policy of treatment priority: given that healthcare typically saves individuals who are at higher *ex ante* risk, then there should be a weighting towards providing healthcare rather than preventive measures (Daniels 2015; Badano 2016).

The extent to which this indicates that a 'high-risk' rather than a population strategy should be followed, is contestable. The 'high-risk' strategy may lead to more cases in total, which when they occur will be of the same severity, than the 'low-risk' strategy. So in practice it will be likely that some mixture of a 'low-risk' and a 'high-risk' strategy will be required. It is also important to notice that (as Chapter 9 explores), health inequalities can themselves be thought as the condensation of *ex ante* risk factors in the realm of the social determinants of health. One crucial reason why poorer people have worse health is due to systemic features that serve to concentrate risk (what Wolff and De-Shalit 2007 describe as corrosive disadvantage), and so the beneficiaries of some types of prevention (just as those in treatment) may be subject to concentrated risk. So, if it is the overall worst off rather than the worst off in health terms who matter most, then this may provide an additional reason to favour population-level 'low-risk' strategies, as Chapter 9 explores.

7.9 Measuring Claims

Hopefully, it should, by this stage, be fairly obvious that it is bizarre to assume—as many philosophers have—that the strength of competing claims can simply be stipulated, and that the 'real' question is how to prioritize between competing claims of a given stipulated strength. As we have attempted to specify more closely how to interpret the elements that have usually been taken to compose the strength of claims, it has become apparent that how to do this is itself highly complex and philosophically controversial; and that it is far from clear that there is a single right answer. Much of the philosophical debate thus risks being pursued with a spurious precision.

Given the variety of ways in which claims could be specified and the lack of consensus on this, the pressing question is how to measure the strength of a claim in a given context. Simply assuming that this can be done and that there are uncontroversial answers to who has the stronger and who has the weaker claim amount to ignoring, rather than solving, the problem of external validity. I shall argue that it would be better to acknowledge that there is a plurality of valid ways of measuring claims, and that how to define and measure claims is itself an important part of the process of clarifying policy goals.

In order to get to this point, we need to step back a little and to think about measurement more generally. We can distinguish between two kinds of broad approaches to measurement: realist and constructivist. On realist approaches to measurement, the thing to be measured is taken to exist wholly separately from the act of measurement: while there may be a history and a progression in human attempts to measure things of this type, progress in measurement of that thing will involve a divestment of particularly human assumptions and perspectives. The development of the International System of Units (SI) is most naturally understood in a realist way: seven base units are defined (for measuring mass, length, time, and so on), which are all based on invariant physical constants, and other measures are defined in terms of these.[21]

Constructivist approaches to measurement take the objects that are measured to be partially constructed through decisions about what to measure, and how to measure it. Many of the key focuses of social policy cannot really be understood as measurable in anything other than a constructivist way. Chapter 9 examines this issue at length through thinking about the measurement of health-related inequalities. I will not anticipate this account here, but instead introduce constructivism more informally through a brief account of crime rates.

What counts as a crime is not fixed as something external to human agency. Which actions are criminalized is something that differs both between different societies, and also changes within the same societies over time. Some offences, such as corporate manslaughter, may be introduced, and others, such as aiding suicide, withdrawn. Even taking the crimes on a particular statute book in a particular state at a particular time as fixed, crime still cannot be treated as measurable in a realist manner. In order for someone to be convicted of a crime, various things need to happen: someone needs to notice that a prima facie crime has occurred and report it to the authorities; police then need to investigate and take the decision to charge the individuals believed to be responsible; and the individuals need to be convicted.

It would be implausible to regard the statistics for any one of these steps alone as 'the' crime rate in a particular locality. Viewed from one perspective, whether

---

[21] With the introduction of the new SI in 2019, the kilogram was redefined in terms of the Planck constant. See Liebisch et al. (2019).

or not a crime occurred is something that can usually only really be settled in a court of law. Outside of the rare case of strict liability offences, it is not actions per se that are criminalized, but actions performed with a particular state of mind or intention (*mens rea*). The state of mind of the person performing an action (along with the nature of the action itself) will jointly determine whether a crime has occurred and, if so, which one. For example, an act of killing is not a crime if it is determined to have been self-defence.

Looked at from another perspective, it is clear that much crime is for various reasons unreported—perhaps because of fear, or shame, or because reporting crime seems like a waste of time due to a lack of faith in the system's ability to bring perpetrators to justice. True rates of, for example, sexual assault have been argued to be very significantly higher than their reported levels. So it would be implausible to take rates of reported crime as 'the' true rate of crime. Relying only on the rates at which individuals are charged or are convicted would be no less problematic, as this would make the crime rate dependent on the efficiency of the police service, or state prosecutors. Crime rates would go down as state officials became worse at their jobs! The net result is that most societies end up adopting a variety of different measures of crime, in the awareness that each measure will be partial in all kinds of ways.

Is measuring the strength of claims to receive a health-related intervention more like measuring mass, or measuring crime? Looking closely at cases in which health systems adjudicate competing claims to receive health interventions in real-world circumstances provides strong reasons to think that constructivism better explains the relevant phenomena than realism. It is common ground that the competing claims theorist needs to specify the feature or features that are relevant to the strength of claims, how to measure these, and how to weight the different features (if there is more than one) to form an overall account of the strength of a person's claim. To take a realist approach to the measurement of claims would be to insist that there is a single right answer to these questions of specification, measurement, and weighting. The constructivist approach would argue that how to measure the strength of claims need not have a single answer. For the constructivist, it will be important to take into account the purposes for which measurement of claims is going to be done, and hence different contexts may make different approaches appropriate.

Nagel's (1979) initial discussion seemed to suggest that the only relevant factor in considering the strength of a claim is how badly off a person is, but as this chapter has discussed, it is increasingly common to take a broader view. For example, Voorhoeve argues that 'the strength of a person's claim is a function both of how much a person could gain in well-being and how badly off he would be without this gain' (Voorhoeve 2014: 69). It is far from clear that policymakers should stop at only two features being relevant to the strength of claims. In so far as the account aims to provide ethical insight into how to adjudicate between competing claims to scarce resources, it is important to notice that other features, such as responsibility, merit, opportunity costs, whether the person is a citizen or an undocumented migrant, and reasonable expectations, are all often taken to be relevant in real-world contexts.

Competing claims theorists do not usually explicitly deny that any such considerations are relevant, but rather aim to forestall such additional complications by stipulating that their scenario is one in which such concerns are not in play and all else is equal. This then raises an obvious challenge about what implications the competing claims account has in cases where everything is *not* equal. Either the competing claims theorist would need to admit that strength of claims thus considered is only one element of what determines fairness in distribution of scarce resources; or they would need to put forward a rather richer account of the elements that compose claims (some of which, for contingent reasons, are not in play in the simpler thought experiments they usually focus on); or they would require an argument for why these other broader elements that are often taken to be relevant to the distribution of scarce resources are not in fact relevant.[22]

There is an additional challenge for any would-be realist account of the measurement of claims. Even simple accounts of features that go into composing claims (such as Nagel's or Voorhoeve's) presuppose that well-being is measurable. In order to determine who is worst off at the point that a decision must be made, and who stands to lose or benefit the most, decision-makers must be able to measure how well- or badly off individuals are. This implies that there is another question about measurement embedded within the overarching question about how to measure claims: how to measure well-being. If the measurement of well-being is better understood in constructivist, rather than realist, terms then it is

---

[22] The ethical context of risk imposition is also often argued to be much richer than the simple models used by competing claims theorists usually acknowledge. For example, Hermansson and Hansson (2007) argue risk management problems can be modelled as having three main parties: (1) those on whom the risk is imposed; (2) those who control the risk, and (3) those who benefit from the risk being taken. On their model, keeping the severity of the harm, and the likelihood of it occurring constant, risks are least ethically problematic when the person or persons imposing the risk also benefit from the risk being run, and are in a position to control the risk. A case of this sort would be if an experienced and knowledgeable mountaineer takes on a challenging ascent. Risks are most problematic where those exposed to the risk neither benefit from it, nor are in a position to control risk exposure. Air pollution provides a good example of this kind of case. Those living in cities are frequently exposed to dangerous levels of PM 2.5 pollution—an exposure that is difficult to control, and usually does not directly benefit those so exposed (Landrigan et al. 2017).

Moreover, as we have seen, accounts of how to measure claims need to take account of the role of risk. Categorizing someone as high or low risk itself depends on assigning them to a reference class. The reference class a person is assigned to will make a significant difference to what their baseline risk rates are taken to be, and how much their risk would be reduced by potential interventions under consideration. However, there is not a single uniquely correct answer to which reference class an individual belongs (Hájek 2007). It has been argued that what is required is an ethically infused account of apt categorization to help inform choices about which categories to adopt for the purposes of public policy (John 2013).

hard to see how the strength of competing claims could be measured in anything other than constructivist terms.

While much philosophical work on the nature of well-being has construed it in a way that would make it apt for realist measurement, it is notable that very few outside of philosophy find such an approach plausible, and that none of the philosophers who presuppose a realist approach to well-being measurement has actually operationalized a way of measuring it. Those seriously involved in measuring well-being tend to be deeply sceptical of the idea that there is or should be a single account of well-being for all human beings in all circumstances (Mitchell and Alexandrova 2020). In fact, different accounts of well-being are designed and validated for different kinds of purposes: poverty reduction, improving the lives of children, or caring for older adults with early stage dementia. Those who construct such measures take care to attend to the specifics of circumstances, as well as the way that measures will be constructed and the purposes for which they will be used (Alexandrova 2017). Overall, while there is nothing incoherent in the idea of a realist approach to the measuring of claims, which has embedded in it a realist approach to measuring well-being, what prima facie philosophical plausibility such an approach has may be due more to a lack of rigour and imagination in thinking through these problems of measurement, than in the availability of a workable realist way of measuring well-being (Mitchell 2018).

7.10 Conclusion

Deciding fairly and transparently which risk reductions to prioritize is crucial for public health policy. The commitments at the heart of the right to public health account—namely that public health policy should fulfil the state's duties under the right to public health, and that all public health activity should be justifiable to individuals—entail that it is important to be able to articulate the implications of public health policies for different groups of individuals, and why policies are fair towards 'losers' as well as 'winners'.

As the analysis of this chapter has shown, there is no simple answer in the abstract to which risks to health matter most that can then be applied by policymakers to give wise guidance in concrete contexts. The philosophical literature has been helpful in distinguishing some dimensions that will be relevant to prioritization decisions, such as priority to the worst off and capacity to benefit. It also allowed us to raise a number difficult questions that all health systems will need to grapple with, such as how to factor in questions about uncertainty, whether and if so how time discounting should be applied to future benefits, and how to compare policies that bring small risk reductions to many against those that bring larger risk reductions to fewer.

This analysis should also have made clear that the philosophical literature is a long way from being able to advance a theory that combines all the relevant elements in a coherent and uncontroversial way, even in the simplified thought experiments through which much of these debates are conducted. The chapter concluded by arguing that in order to be able to provide guidance that could plausibly have external validity, such ethical discussions need to shift away from the abstract question of how to prioritize between different claims of given stipulated strengths, towards how to actually measure the strength of claims in the context of public health policy. Thinking in more detail about how claims could be measured gave us reason to doubt that the best theory of measurement for claims will be a realist, rather than a constructivist one. If the strength of claims is best understood in constructivist terms, then it is implausible to think that there is a single and unambiguous answer to how best to measure claims. There will be multiple possible ways of so doing, depending on the context and the purpose of the policy to which this measurement will contribute. Part III takes forward these questions about prioritization through close analysis of three important domains of public health policy.

PART III

STRUCTURAL JUSTICE

Introduction to Part III

The promise of abstraction is that ethical problems can be simplified by stripping away inessential features such as context and history without losing anything of importance, and that analysing simpler models of ethical problems will make real-world decision-making both easier and more accurate. Part I argued that this idea is seductive, but illusory: the unsolved, and probably insoluble, challenge is how to bridge the gap between establishing internal validity in the simplified and purified context, and external validity in the messier and more complex real-world environment in which policy problems arise and subsist.

In order to stand a good chance of success, public policy needs to take account not just of the causal complexity of systems at different levels, but also the pluralism and social construction of the values that are invoked setting goals, and measuring success. Taking complexity seriously means that policymakers need to take account of performativity—the ways in which these questions about value are interpreted by actors within the system. A top-down model in which interventions are designed and ethically appraised in the abstract before being imposed without any input from ground-level is likely to lead not only to policy failure (in that the policy fails to achieve its goals), but also to ethics failure (in that the policy ends up being much more ethically problematic in practice than was anticipated).

Knowing that systems are complex and interrelated, and that interventions need to take into account citizens' values alerts decision-makers to certain kinds of problems of external validity, but it does not itself amount to a positive view of what is worth pursuing in public health, and how to approach making trade-offs. The chapters in Part II advanced an ethical framework that can guide policymakers in finding solutions to specific problems that arise in their own contexts.

This framework comprises four main elements. First, a commitment to public health as a right of citizens. Governments' duties are not restricted to intervening only to avoid harm to others, but also include positive duties to reduce population-level health risks, and to promote health. Where a government (or other agents, such as commercial companies) fall short of doing what is required, then it is appropriate to ask whether individuals' right to public health has been violated.

Second, public health interventions must be justifiable *to* the individuals affected by them. Health improvements at a population level—even those that would be predicted to significantly reduce mortality—are unlikely to be worth having if it would be worse *ex ante* from the perspective of a representative individual to have the package of the slight risk reduction and its accompanying inconvenience, than not to have the intervention at all.

Third, while policy interventions need to be justifiable *to* individuals, this does not imply that they should be targeted *at individuals*. Individualistic justification is not only compatible with, but will often require, interventions that are targeted upstream from individuals, and which have their effects only indirectly. Intervening at a broader social level to improve health will often be easier to justify to individuals than interfering with individuals' lives directly, given that it is easier for upstream interventions to bring risk reductions without disproportionate inconvenience than it is for interventions that require individual-level behaviour change.

Fourth, promoting health in the narrow biomedical sense is only one of a number of legitimate goals for governments, and, moreover, many individuals will understandably treat biomedical health as only one of the things that matter to them. There will be reasonable disagreement both amongst citizens and amongst policymakers about the appropriate weight to give to the protection and promotion of health. Public health will often need to seek synergies with other goals, rather than simply asserting its own priority.

The framework aims to clarify the ethical features of a range of choices that need to be made about goals and means in public health, rather than attempting to settle definitively questions that are subject to reasonable disagreement. Thus, Chapter 5 suggested that at least four separate variables matter in justifying state activity that would curtail liberty or autonomy in the name of improving health: the size of the health benefit to be obtained; the level of public support; the extent to which the policy involves interfering with autonomous choices; and the extent to which it involves interfering with liberty—without requiring that these are the only relevant variables, or requiring anything prescriptive about the relative weights that should be placed on them. Chapter 7 examined a number of choices that need to be made in deciding which health needs are most important to meet—clarifying the different dimensions of priority to the worst off, capacity to benefit, time, and risk concentration, and the costs and benefits of taking different positions.

Notwithstanding this framework, some readers—particularly those who work as policymakers within public health—may find this advice too high level and lacking in specificity. From the ground level, it is of limited help to know that things are complex and that pluralism is required, without also having some guidance as to *how* to make ethically sound decisions in specific areas of public health, given this complexity and need for pluralism.

Part III closes this gap by examining three spheres of influence that are of relevance to public health, and to public policy more broadly. In each case, public policy will need to make some choices about how to draw lines and where to place emphasis. The approach that states take in each of these spheres—responsibility, equality, and contagion—has a profound effect on the conditions in which individuals either thrive or deteriorate. The core challenges posed by these spheres for public health are: (1) how to make use of judgements of responsibility, and whom to hold responsible; (2) how to specify the goal of health equity and how to pursue it; and (3) how to respond to the fact that most health risks are either contagious or can be amplified by socially mediated networks of causes. Approaching these challenges from a systemic perspective, the key question is not so much just how to do justice to individuals on a momentary basis, but how to structure institutions, networks, and incentives in ways that maintain and strengthen social justice over time. I describe this as taking a structural justice approach.[1]

Public health policy by its nature requires a reckoning about how to distribute the burdens and benefits of reducing health-related risks. The idea of personal responsibility has recently ascended in popularity in policy discussions of how to do so. Chapter 8 critically examines the idea of responsibility more broadly, arguing that which aspects of an individual's condition they can reasonably be held responsible for is an ethical, rather than a factual, question and that there is good reason to think that the answer to this ethical question should be based on the values that the health system is aiming to promote or respect, rather than treating personal responsibility as an ethical requirement that places extrinsic limits on health system design.

The right to public health account provides some guidance on how to think about responsibility in health policy, though it will also leave a degree of play for contextual determination. On the right to public health account, a health system's answer to the question of whom to hold accountable, and how to do so, needs first of all to be framed within the context of the right to public health. In so far as claims of irresponsibility can be fairly levelled, these should in the first instance be directed towards those who violate the right to public health, either through government or corporate agency, rather than at isolated individuals. Health systems should aim to expand and protect individuals' effective ability to *take responsibility* for their health, but it will not usually be the case that *holding these individuals accountable* (or threatening to do so) will be a particularly effective means of so doing. Acknowledging the importance of individuals taking responsibility for their own health is consistent with a resolute resistance to blaming or

---

[1] As was explained in Chapter 1.5, the analysis in this book aims to apply to middle- or high-income democratic states, which have an effective government, and a commitment to treating their citizens as equals.

otherwise holding to account those who, for one reason or another, do not do so effectively.

Chapter 9 examines how health systems should measure, and respond to, health-related inequalities. Health equity is often taken to be a core goal of public health, but what exactly health equity requires is more difficult to specify, for two reasons. First, there are indefinitely many health-related variables that can be measured, and variation in each of these variables can be measured in a number of different ways. Second, given the systemic interconnections between variables, making a situation more equal in some respects will make it less equal in other respects. The chapter argues that a number of existing philosophical approaches are too simple: they tend to assume that there will be only one type of health inequality that really matters, and also tend to assume that it is inequality per se that is the problem rather than features which accompany inequality. The chapter argues that it is better to adopt a pluralist approach to health equity measurement. What measures of health equity need to be accountable to is the lived experience of individuals' lives and the ways in which power impacts on these. Reflection on the deepest and most resilient causes of health-related inequalities shows that they are often the result of intersecting structural concentrations of power— structures that it is vital, but very difficult, to break up.

Chapter 10 examines the idea of contagion—of risk magnification and modulation through networks. While this is most familiar within public health in the context of communicable disease, adopting a complex systems approach makes salient that infectious pathogens are only one of a number of ways in which networks matter for public health: social norms, food cultures, and attitudes towards body shape are all shared and amplified through social networks and are thus contagious. The chapter examines three case studies, which each raise different questions about the interplay of causal complexity, performativity, and policy-making: vaccination policy, drug resistant infections, and disease eradication. In vaccination policy, achieving herd immunity is often crucial, but attempts to do this are heavily dependent on public trust. Drug-resistant infections arise, among other causes, through the inevitable impact of natural selection, and so require a shift towards an ecological perspective on disease. Finally, the possibility of disease eradication poses important questions about when and how to ensure that susceptible health threats are systematically and permanently removed from the environment.

8

Responsibility

8.1 Introduction

Choices about which public health policies to pursue are nearly always also choices about the distribution of risks and benefits. Policymakers must not only make trade-offs between values such as cost-effectiveness and priority to the worst off, but also determine how responsibility for reducing health-related risks should be distributed between state institutions, third parties such as commercial companies or non-governmental organizations (NGOs), and individual citizens.

The argument of Part II aimed to establish that the reduction of health-related risks is a core responsibility of governments, and that if a government falls well below the required standard of health protection and health promotion, then it violates the rights of its citizens. In establishing a robust role for public health, Chapter 6 introduced the idea of the Neglectful State as one that fails in its responsibility to protect and promote health. The argument acknowledged, but did not explore, that states are not the only entities that can violate the right to public health. It also left unexplored the relationship between the responsibilities that states have to protect and promote health, and any responsibility that individuals have to protect their own health. This chapter takes up these questions.

As we get underway, it is important to remind ourselves of some of the points about causation that we have covered elsewhere in the book, as these will make a difference to how to think about questions of responsibility. Chapter 2 explored some of the ways in which accounts of causation that can be relied on for externally valid policy interventions need to be more nuanced and sophisticated than common-sense ideas of causation often are. Even in relatively simple cases such as a fire starting in a warehouse, what is picked out as 'the' cause (such as a short-circuit) is actually part of a set of causal conditions, and the result would not have materialized without these support factors.

Which feature is picked out as 'the' cause is often more an expression of the interests and normative assumptions of the person making the causal attribution, than an identification of a causal factor that was on its own either necessary or sufficient for the effect to occur. Judgements about causal responsibility may also incorporate assumptions about what different actors would reasonably be expected to do. For example, if the lifeguard on duty failed to notice that a child was drowning, this is much more likely to be cited as a cause of the ensuing death than if a bystander fails to notice this.

Complex systems greatly magnify these difficulties and ambiguities in identifying causes, given the multiple feedback loops, frequency of delayed and nonlinear effects, and the many points at which interventions (or failures to intervene) could occur. In complex systems, questions about the relationship between identification of causes, understanding of role-based or other ethical obligations, and ethical responsibility for harms that transpire will frequently be controversial. First, there will be different ways of mapping the causal structure of the system, and these different ways will emphasize the role of different agents and processes to different degrees, and thereby make some rather than other assignments of ethical responsibility more plausible. Second, where different processes interact in a way that increases likelihood of harm to some, many or all of these individual actors will be able to say that their part in whatever harm results is negligible, even if the overall effects of lots of people performing similar actions is hundreds or even thousands of deaths per year.

8.2 Assigning Duties Under the Right to Public Health

The right to public health implies that there are duties on other parties to respect this right, and that if those duties are not fulfilled, then the right is violated. So the right to public health (and other rights where appropriate) gives us an important baseline for thinking about ethical responsibility in public health. As with any rights claim, specifying it will require further normative work to answer the following questions. First, what is the right a right *to*? Second, who are the duty bearers of the right? What does acting in accordance with the right require from its different duty bearers?

What the right to public health is a right to was answered in outline in Part II, and is fleshed out in the concrete cases of health-related inequalities and communicable disease in Chapters 9 and 10. To summarize, the right to public health is entailed by a more general right to health; and it implies a right that proportionate and cost-effective measures to reduce health threats from the environment be taken. In specifying what is required under the right, it is important that public health policies are pursued in a way that is justifiable *to* individuals—where doing so involves treating them as equals, and also acting in a way that takes proper account of the importance of other rights and goods such as liberty and autonomy.

Chapter 6 established that the state, at least, is a duty bearer under the right—as it is both in a good position to reduce health-related risks, and also has a duty to promote the common good. In fact, any agent that could affect the profile of health risks that citizens are subject to is a duty bearer under the right to public health. Claim rights in general have this feature of placing duties on the world at large: for example, my right not to be tortured entails that everyone else has a duty not to torture me. However, there is a difference between an agent being a duty bearer in this very broad sense, and this fact impinging in a significant way on what this agent can and cannot do on a day-to-day basis. So the crucial question is not so much can there be universal duties that follow from the existence of the right to public health (as this follows automatically), but rather how should the right to public health impinge on the actions of institutions, commercial entities and individuals?

There will be some clear cases in which the actions of nonstate agents such as private citizens or companies violate the right to public health. For example, if a private individual introduces a pollutant into a city's water supply, or if a company negligently allows a poisonous gas to escape from an insecticide factory, then both may plausibly violate the right to public health. Going beyond the obvious cases requires a combination of careful empirical and normative work. It will rarely be the case that there is only a single point at which intervention can be made, or only a single factor that is responsible for a large-scale population health harm.

The right to public health account argues that if the state takes a laissez-faire attitude, then it is unlikely to lead to a distribution of risk that is optimal. For example, it may lead to risks being multiplied for those who are already worst off; or people needing to take expensive and ineffective measures to guard against risks that could have been prevented at source. The crucial question is a more specific one: how much responsibility for control of particular kinds of health-related risk should be placed on the shoulders of private individuals, and how much elsewhere in the system—whether on state or nonstate actors?

Air pollution provides a good way into the basic contours of the challenge. Suppose that a city currently has levels of air pollution that are significantly outside those recommended by internationally agreed standards, and that this is causing significant harms to health, leading both to increased early deaths and a significant number of chronic illnesses. One thing that the state could do would be simply to let this situation happen—and not to attempt to regulate. If a state does this, we can further suppose that some individuals will be able to mitigate these risks down to a level that they find acceptable, by, for example, paying attention to public reports of pollution levels and staying inside when advised to do so, avoiding outdoor exercise when pollution levels are high, installing an air filter in their home, or moving house.

Framing a response to the ill health caused by the air pollution in laissez-faire terms would be to fail to take account of two ethically important features. First, it would ignore the fact that levels of air pollution are much higher than is recommended by international standards. So adopting this approach would be to fail to engage with the larger question of *why* levels of air pollution are so high, and how to reduce them. Proponents of the right to public health will insist that there will be some cases where some individuals *could* take steps to mitigate a health risk as individuals, but they *should not have to*, because the risk is not of a type or magnitude that any citizen should be exposed to.

Second, the ability to act on the basis of advice about air pollution will not be uniformly distributed in the population. Some jobs will require outside working in areas of high pollution; and some pollution-reducing measures (such as installation of air filters in the home or moving to an area of lower pollution) will be out of range for the vast majority. The fact that it would, in some fairly broad sense be *possible* for individuals to mitigate a risk does not imply that it would be feasible for them to do so.

There are a very wide range of individuals, institutions, and corporations whose behaviour is causally relevant to the air pollution-related risks that citizens face. Judgements of responsibility, and potentially blame, could be made about any of the individuals, institutions, or corporations that are in a position to modulate these risks—whether altering their likelihood or severity, or the degree of control that citizens have over their risk exposure. Activities that could be assessed in responsibility terms would include politicians' role in establishing a regulatory framework for air pollution; public health officials' role in enforcing existing provisions; the design choices of manufacturers of items such as diesel cars or wood-burning stoves that cause air pollution; business owners' choices about use of fossil fuels and polluting chemicals; car drivers' choices about their type of car and the amount they drive; as well as what individuals do or do not do to protect themselves against risk factors to their health.

The fact that a particular agent, institution, or corporation *could* reduce a health risk does not yet show that they ought to, or should be held accountable for not doing so. Determining where responsibility should lie for reducing particular health-related risks thus requires not only an empirical inquiry into the causal factors that are acting to worsen a particular health outcome and the causal factors that could be mobilized to improve it (these need not be the same), but in addition a normative enquiry into how ethical responsibility should be distributed among these potential contenders.[2]

The right to health (or public health) does not by itself imply that the individuals whose health is at stake ought to take steps to improve their health, or that they have done anything wrong if they fail to do so. While, as will be discussed below, some have argued that there are good policy reasons for thinking that it is justifiable (or even ethically required) for states to condition availability of publicly

---

[2] One implication of the ethical framework advanced in Part II is that while public health should be an important goal of states, it is not the only goal that matters. So the fundamental challenge for states is to work out how the right to public health is to be protected alongside other rights and goals. As the framework advanced in Chapter 5 pointed out, the right to public health account will favour ways of reducing risk that minimize impact on individuals' liberty and autonomy—and that this will usually be better done by acting in a way that shapes environments, rather than requiring individuals to change behaviour. I consider what this entails in more detail in the case of communicable disease in Chapter 10.

funded health services on the basis of judgements about personal responsibility, such a view does not follow directly from the right to public health.

8.3 Substantive Responsibility

So far, I have been relying on an intuitive conception of responsibility, but we now need to sharpen this up by making some distinctions. First, we can distinguish causal and moral responsibility. Judgements about causal responsibility are based on a determination of the extent of the causal contribution that different factors made to an outcome. An agent can be causally responsible for a particular outcome without being morally responsible. In an inquest into a death after a traffic collision, it may be common ground that the pedestrian died as a result of being struck by a particular car, and to that extent it may be agreed that the car (and its driver) was causally responsible for the death. Notwithstanding the clear causal responsibility, there will be circumstances where it is judged that there is no moral responsibility—such as if the pedestrian stepped right in front of the car without any warning.[3]

This chapter focuses on moral responsibility. Scanlon (1998) makes a helpful distinction between two modes of moral responsibility: what he describes as responsibility as attributability, and substantive responsibility. An action (or inaction) is *attributable* if it can be 'attributed to an agent in the way that is required in order for it to be a basis for moral appraisal' (Scanlon 1998: 248). Actions that are performed voluntarily, by competent agents who are aware of relevant features of what they are doing, are attributable.[4] To hold someone substantively responsible is to make a judgement that their behaviour makes a difference to how others should treat them. While some ways of talking about responsibility imply that the conditions under which conduct is attributable to an agent, and the conditions under which the agent performing it is substantively responsible are the same, it is important to notice that this represents a decision rather than a conceptual entailment (Scanlon 1998: 249).[5]

---

[3] The agent is likely to still *feel* some degree of responsibility in such circumstances—even though both friends and state officials will make clear that the driver is not to blame for what happened, and the driver would not be held legally liable. See the discussions of agent regret and of moral luck in Williams (1981) for further on this point.

[4] There will be a range of other cases where there is some room for disagreement about whether an action is attributable—as when someone chooses as a result of a misleading presentation of the facts, or a decision is taken as a result of a threat, or the person taking it is very drunk. While such cases raise interesting questions, I leave them on one side.

[5] In a helpful discussion of how to interpret Scanlon's views in the context of public policy, Mounk (2017) argues that it would be more intuitive to explain what Scanlon describes as moral attributability as 'responsibility', and what Scanlon describes as substantive responsibility as 'accountability'. Thus, within Mounk's terminology, the main point of Scanlon's distinction would be: 'even if we decide that some individuals are in fact responsible for a particular action or outcome, a question remains as to

There are a range of current practices within public policy that deliberately separate attributive and substantive responsibility. Legislation that allows for individuals to file for bankruptcy provides an interesting example of case where policymakers have explicitly decided on ethical or policy grounds *not* to hold agents substantively responsible for some things for which they are attributively responsible. Declaring bankruptcy allows individuals to start again and for the slate to be wiped clean—irrespective of whether the circumstances that led to them being insolvent were decisions for which they were attributively responsible. Such legal provisions could be defended ethically by reference to the unconscionability of the kind of regime that preceded it, in which the insolvent could be placed in a debtors' prison or a workhouse until their debts were paid off.[6] It could also be defended through a commitment to supporting entrepreneurialism and risk-taking in the economy—and a correlative requirement to make financial failure less costly to the individual. Either way, most successful economies have legal regimes that allow individuals to start again even after insolvency caused by their own failure or foolishness.[7]

Conversely, policymakers can decide to hold agents substantively responsible for particular outcomes, regardless of whether they were attributively responsible. Strict liability offences provide a good example. Usually criminal offences require both an *actus reus* (guilty act) and a *mens rea* (guilty mind), but strict liability offences punish particular acts regardless of the mental state of the perpetrator. Within the USA, several states make statutory rape a strict liability offence: even if the perpetrator reasonably believed the other party to be over the age of consent, and the other party did in fact give consent, this provides no exculpation. Strict liability is ethically controversial, just in virtue of separating liability for punishment from any requirement for *mens rea*, but it has been defended as ethically justified in cases where it provides an effective means of ensuring that individuals or companies take precautions that they otherwise may not have taken, and in order to ensure systemic change.[8]

In systems that function safely and effectively, there are likely to be various forms of redundancy in the ways in which risks are governed and regulated. If one person or one layer misses a problem, then it will usually be caught elsewhere.

---

whether or not we should hold them accountable by changing our treatment of them' (Mounk 2017: 157). While I have found Mounk's analysis helpful, I have retained Scanlon's terminology, in deference to its wide use within the literature. I have benefited also from Walker's (2019: ch. 6) discussion, which interprets Mounk's analysis in the light of the treatment of chronic disease.

[6] A situation memorably described by Charles Dickens in his *Little Dorritt*.

[7] The lived experience of bankruptcy—particularly in a society that is shifting towards greater inequality, and a more judgemental approach to individual responsibility—may be different from that intended by the framers of the legislation. In a historical analysis of US bankruptcy data from 1977 to 2016, Sousa (2018) shows that as rates of bankruptcy increased, feelings of stigma around bankruptcy also increased, and links this connection to the rise of individualistic models of personal responsibility.

[8] For discussion of the ethical issues raised by strict liability, see Simester (2005).

Analysis of large-scale failures within public services shows that they are very rarely down solely to a single individual, but are more usually due to a cascading series of failures at different levels. James Reason makes a helpful and often quoted analogy to Swiss cheese. Each regulatory layer of checking will have some holes in it; and in order for catastrophe to happen the failure event needs to travel through holes in various different layers (Reason 1995).

In areas in which there is a desire systematically to reduce errors—paradigmatically in cases such as airline safety and fire safety—researchers and practitioners have argued for a shift away from individualized blame as an approach to substantive responsibility, and towards adoption of systems approaches. Eileen Munro, who led a groundbreaking report into a mistake in social care, explains as follows:

> Disasters are rarely found to happen because of one major mistake by one grossly incompetent worker but to be the result of a system operating with a chronic pattern of small errors or omissions, most of which have no serious adverse effect but which, on one tragic occasion, come together to lead to a major accident.... When a traditional investigation identifies human error as a cause, it is assumed that the person who erred 'could have acted differently'— that he or she can be held responsible for omitting a crucial step or misinterpreting a vital piece of information. The systems approach has a more complicated picture of causality. The human operator is only one factor; the final outcome is a product of the interaction of the individual with the rest of the system.
>
> (Munro 2005: 534)

Thus, the systems approach to thinking about major adverse events emphasizes the need to design systems for safety: to build cultures around safe practices, and to ensure that attention is paid to near misses (and that there are structures requiring near misses to be reported and investigated). The account of substantive responsibility relied on is one that is deeply shaped by the goal of systemically reducing risk.

In other cases, a systems analysis may suggest that particular needs are being neglected because they fall in a zone where it is unclear which institution has responsibility for meeting them.[9] If so, the responsibilities of one of the relevant bodies could be expanded to definitely encompass the neglected needs; or shared goals be set between the institutions to avoid the problem of ambiguous responsibility. Just as in the case of avoiding disasters, altering the assignment of

---

[9] As Williams (2008: 467) puts it, 'responsibilities are always liable to fall through the gaps; changing realities are always liable to disrupt existing divisions of responsibility; actual powers may be at some distance from notional responsibilities. Or in other words, if everyone merely "does their job," organizational irresponsibility may still result. In this situation we must all pick up the pieces, look out for unmet responsibilities—without falling into insubordination or otherwise infringing on others' spheres of responsibility.'

substantive responsibilities will also alter the causal factors that are in play, weakening some causal pathways and strengthening others.

While the purpose and value of the distinction between attributive and substantive responsibility should by now be fairly clear, unfortunately much of the debate both at a popular level and in the academic literature ignores it. This has led at times to a slightly cross-purpose discussion, in which some who wish to deny that judgements of substantive responsibility should be used in health prioritization decisions sometimes feel the need to argue that behaviours that lead to health-related disadvantages never, or very rarely, meet those standards for attributive responsibility. As we shall see, the right to public health account will recommend separating substantive responsibility from attributive responsibility, but need not deny that people have attributive responsibility for their health-related behaviours.

8.4 The Social Democratic Vision of Responsibility, and Its Decline

Social democratic welfare states of the kind that flourished in the 1950s to the 1970s in Western Europe, and canonically in Scandinavia, tended to adopt an approach to substantive responsibility that departs very considerably from attributive responsibility. They did so deliberately and in order to ensure that the government and employers shoulder a great deal of substantive responsibility when it comes to public health, while individual citizens are asked to shoulder much less.

Governments that were keen to make systemic changes towards a safer society adopted a strict liability approach to workplace health and safety. Under such approaches, an employer could be punished for harm that comes to their employees even if there was nothing they could have reasonably done to prevent the harm that resulted on a particular occasion—thus strongly encouraging the creation of a safety culture. Health systems also deliberately excluded questions about how individuals came to be ill from consideration of their eligibility for treatment, or the priority that they should receive. Even if a particular injury was self-caused and the person who injured themselves was aware of the dangerous situation they were exposing themselves to, the system treated such facts as irrelevant.

Rawls's theory of justice provides a powerful articulation of the kind of ethical view that underlies the social democratic model. Rawls confines himself to discussion of how to set up the fundamental social structures correctly (what Rawls called the 'basic structure' of society), and argues that judgements of personal responsibility or of desert should play only a minor and subsidiary role in these discussions. Rawls approached the problem of the design of just institutions via his veil of ignorance thought experiment, asking what principles would be chosen by free and equal persons who had no idea what role or position they were going to occupy in society—whether man or woman, rich or poor, a member of a majority group or of a minority. On Rawls's view, given their deliberative situation, the parties to the original position would start from an initial assumption of equal shares, and assume that departures from equality need to be justified; inequalities could not simply be assumed to be fair (Rawls 1999: 130).

Rawls considers two main kinds of argument for departure from equality: arguments from desert and argument from incentives. Arguments from desert would provide reasons for thinking there is something about the moral qualities of individuals' actions and achievements that makes the resulting inequalities intrinsically desirable. Arguments from incentives would provide reasons for thinking that the total effects on a social system will be superior if certain kinds of inequalities are tolerated, even if the individuals who receive the incentives would not have deserved them prior to the setup of institutions. Arguments from incentives thus justify inequalities as a by-product of the pursuit of other legitimate goals, rather than as following from claims that are valid independently of any social structure.

Rawls (1999: 273) acknowledges that there is 'a tendency for common sense to suppose that income and wealth, and the good things in life generally, should be distributed according to moral desert' but he denies that such principles would be chosen behind the veil of ignorance. Crucial to his argument is that no one can deserve something that they had no capacity to influence. So, in order for the desert theorist to make the case that there are deserved inequalities, it would be important to distinguish those features of an individual's life and choices that could appropriately be the basis of a desert claim, and those that cannot.

Rawls argues that individuals cannot deserve (and, although he does not use this language, cannot be attributively responsible for) the circumstances of their birth; the social class into which they were born and develop before they reached the age of reason; their native endowments (such as talents or handicaps that they were born with); any talents that are developed before they reach the age of reason; or the good or bad luck over the course of a life (such as illness and accident; regional economic decline). Importantly, Rawls also argues that these features for which individuals are not attributively responsible also have profound effects on the other features that are usually thought to be the core of our abilities to make responsible decisions: 'Even the willingness to make an effort, to try, and so to be deserving in the ordinary sense is itself dependent upon happy family and social circumstances' (Rawls 1999: 64).

For these reasons, Rawls thought that it would be a significant mistake to incorporate a pre-institutional account of moral desert into the basic structure of society. However, he did think it important to satisfy legitimate expectations based on the actual existing social institutions (Rawls 1999: 273). Although the society guided by Rawls's principles of justice will allow inequalities of income and wealth, it will do so only because of the incentive effects that some levels of inequality create. Doing so will create a larger social product to be shared, and because of this it will allow the creation of a society in which the worst off are better off than they would have been under strict equality.

Both social democracy and Rawls's approach to justice thus incorporate what Mounk (2017) describes as an institutional account of responsibility, which starts from the nature of the institutions that a society is trying to build and maintain, and uses this to design an account of substantive responsibility. On an institutional approach, it is quite possible to take the view that properly fulfilling the purposes of particular institutions and services requires *not* making judgements about whether citizens who want to access a particular service are or are not attributively responsible for their need. Indeed, the provision of universal benefits without means testing is a central plank of the social democratic approach.

One thing that Rawls focuses on less in *A Theory of Justice*, but which came to occupy him more in his later work, was the ways in which his idealized vision of society could become undone as a result of rising inequality—particularly if increasing wealth inequality went hand-in-hand with increased control over political processes by the wealthy. In the real world there is often a drift between an understanding that a scheme of rewards has been set up to provide incentives, and a belief that these incentives are deserved.

The progress of intellectual property rights talk provides a powerful example. As intellectual property has become an ever more significant source of income and wealth, there has been a shift from a position in which legal rights were explicitly set up in order to provide incentives, to one in which legal rights have come to be seen by many as a response to pre-institutional desert (Wilson 2009a). Waldron explains this process as follows:

> Incentives work by conferring benefits on those whose activity we are trying to encourage. Such a benefit may be seen as a reward for their efforts. Rewards are what we characteristically provide for moral desert; we reward the deserving and penalize the undeserving. Therefore, authors deserve the intellectual property rights that are secured to them in the name of social policy. The thought moves from *encouragement* to *incentive* to *benefit* to *reward* to *desert*, so that something which starts off as a matter of desirable social policy ends up entrenched in an image of moral entitlement.   (Waldron 1992: 851)

This kind of process of creeping transformation of incentives into deserved inequalities has important implications for the stability of Rawlsian and social democratic-style societies over time. To the extent that the incentives to the better off, which are initially intended to ensure a larger overall social product, come to be interpreted as deserved by the wealthy, then the cohesiveness of society will be undermined.

Indeed, the social democratic project, if not dead, is widely perceived to be in deep trouble even by those who are most sympathetic to its aims—undone by a constellation of interrelated factors: a significant rise in wealth inequality; a shift from a universal to a 'safety net' model of the welfare state; an increase in diversity of populations within countries; and a shift to a more judgemental model of responsibility, which aligns models of substantive responsibility much more closely with attributive responsibility.[10]

The heyday of the social democratic project coincided with the Bretton Woods system (1945–71), which imposed stringent capital controls in the world's major economies, and severely restricted capital flows across borders. This enabled income tax rates to be significantly higher than at present: for example, in 1963 the top marginal rate of income tax was 91 per cent in the USA and 89 per cent in the UK. Very high rates of marginal taxation at the top level meant that levels of wealth inequality reduced—both as a result of the tax take, and also the fact of the very high marginal tax rates itself reduced the pressure to increase wages. The decline of the Bretton Woods system saw an end to capital controls, and in the wake of this came international tax competition—and a growing belief amongst policymakers that income and corporation taxes needed to be reduced in order to avoid capital flight.

By the late 1970s, the post-war consensus that had supported the social democratic project began to fall apart. Within the USA and the UK, the Reagan and Thatcher governments inaugurated processes of deregulation and privatization, and significantly weakened their commitment to the idea of redistributive taxation—with many other countries travelling a similar path.[11] By 1989, the top marginal rate of income tax had fallen to 28 per cent in the USA and 40 per cent in the UK, before rising slightly to 35 per cent (USA) and 45 per cent (UK) at the time of writing. Rates of corporation tax also declined sharply during the same time period (see Piketty 2014 for data and analysis). The decline of capital controls also made possible aggressive tax avoidance through tax havens and the use of complex international ownership regimes, which have allowed the super-rich to further reduce their tax liabilities.

Moreover, since the 1980s, the global economy has also transitioned from one in which the predominant source of wealth was physical goods, to one in which the predominant source of wealth is intangible goods such as intellectual property, brands, and media platforms. Markets for intangible goods exhibit features

---

[10] Two of the most astute political commentators on this process are Judt (2009, 2011) and Varoufakis (2016). Mounk (2017) and Young (2013) provide an excellent account of the elements of this story that relate to responsibility.

[11] Pinochet's Chile provided a laboratory for some of these ideas. See Letelier (1976) for an early critique.

such as network effects, high up-front costs but low marginal costs, and customers getting used to particular services and products, which together make for what Arthur (1989) describes as an economy of increasing returns—namely, one in which companies that get ahead will tend to increase their lead. Where there are increasing returns, there are significant path dependencies and markets will tend towards monopoly or at best oligopoly without firm regulation and governance. Of course, the ability for national governments to impose such measures was itself significantly weakened by the rise of the capital mobility that allows multinational companies easily to relocate to territories with favourable regulatory regimes.[12]

The social democratic welfare state was conceived of as a universal raiser of standards. The middle class 'received in many cases the same welfare assistance and services as the poor: free education, cheap or free medical treatment, public pensions, and the like. In consequence, the European middle class found itself by the 1960s with far greater disposable incomes than ever before, with so many of life's necessities prepaid in tax' (Judt 2009). As the 1980s wore on, and especially in response to changes made in the USA by Ronald Reagan and the UK by Margaret Thatcher, the political consensus on the welfare state shifted from the universalistic vision to that of a safety net for those who really needed it. The move away from a universal model brought with it a greater perceived need for testing and conditionality, to ensure that no one who was not eligible was able to claim benefits.

This shift in the political justification for the welfare state coincided with these same states seeing much increased immigration, and so the underlying idea of risk pooling and solidarity that had been taken for granted in social democracy came under pressure from this direction too. Ideas of solidarity within the welfare state had previously been able to be based on common national heritage, but such stories no longer seemed as potent in unifying and justifying solidarity in a multi-cultural society. To the extent that such stories did retain their potency, they were as likely to lead to a solidarity of exclusion that saw newcomers as a common threat to be resisted, as a solidarity of inclusion that expanded and rethought the boundaries of the 'we'.

The final element in the decline of social democracy was a shift in the discourse of responsibility away from thinking of responsibility in terms of duties to others and a duty to support the common good, to a vision that emphasized atomistic self-reliance. As Mounk puts it:

Whereas responsibility had once appealed to a citizen's concern for others, it now primarily refers to the obligation to look after oneself...Its basic tenet is

---

[12] As I discuss in Wilson (2020), these changes are largely responsible for the rise of the number of ultra-wealthy like Bill Gates, who have transformed the global health landscape through their philanthropy.

clear: all citizens have an obligation to look after themselves. Though they can expect society to help them if they end up in need for reasons beyond their control, such as accident or an inborn disability, this form of public solidarity is tied to stringent conditions.   (Mounk 2017: 36)

Each of these elements is interlinked, and tend to reinforce one another. Increasing inequality tends to undermine the sense of social solidarity that comes from all, whether rich or poor, thinking of themselves as living in the same social world. Not only do the rich and the poor come to have very different kinds of experiences, it will also become increasingly likely that the rich no longer have to rely on the same public services as the poor. To the extent that public services come to be seen not as universal, but as something for the poor, this opens the door to more judgemental and more conditionalized policies. Policymakers are much less likely to allow humiliating requirements to be placed on services that they and those in their social circles need to access, than on ones that they think of as for the benefit of others.

Looked at from another perspective, the reduction in social solidarity is itself a driver for a more individualistic model of responsibility, in so far as citizens become less willing to trust others and contribute to solidaristic schemes to benefit them. In turn, both the loss in strength of solidarity, and the rise of individualistic models of responsibility make it more difficult to enact the kinds of legislative and policy changes necessary to significantly reduce income and wealth inequalities—given that the individualistic model of responsibility makes people more likely to think that the wealthy deserve their riches, and that the poor have failed to help themselves. While the moral vision that animated social democracy remains attractive from the perspective of the right to public health, there is no easy path back to the social democratic society.

8.5 Luck Egalitarianism

The sense that the political ground had shifted led to a change of direction within liberal egalitarian political philosophy. By the 1980s, the kind of society that Rawls defended seemed to be wide open to the critique that it was insufficiently attentive to personal responsibility. Rawls's difference principle, which was supposed to govern the extent of permissible inequalities in income and wealth, requires only that social and economic inequalities 'be to the greatest advantage of the least advantaged members of society' (Rawls 1999), without giving any consideration to how the worst off come to be in that position.

Ronald Dworkin argued that the difference principle misses an important element of fairness, namely that what individuals receive as a matter of justice should be responsive to their voluntary choices (Dworkin 1981a, 1981b). In a much discussed thought experiment which aims to motivate this claim, Kymlicka (2002: 70–6) asks us to imagine two people who start from initial equality—including equal talents, social background, and equal resources. One of them is interested only in playing tennis, and uses his land to build a tennis court. He works short hours, and spends the rest of his time improving his backhand. The other uses her plot of land to grow vegetables for herself, and to sell to others. She puts in long hours at work. Without some form of redistribution, the gardener would quickly come to have more resources than the tennis player. If we assume that the tennis player would then become one of the worst off, this uncovers a lacuna within Rawls's account: he had not thought to distinguish between those who are amongst the worst off because of circumstances beyond their control, and those who are badly off due to their own voluntary choices. It seems unfair, Kymlicka argues, if the gardener ends up subsidizing the tennis player's lifestyle: 'Treating people with equal concern requires that people pay for the costs of their own choices' (Kymlicka 2002: 74).

Cohen argued that a core strength of the kind of kind approach that Dworkin proposed was that it allowed liberal egalitarians to provide a dialectical response to right-wing critiques of social democracy, and thus to incorporate and defuse 'the most powerful idea in the arsenal of the anti-egalitarian right: the idea of choice and responsibility' (Cohen 1989: 933). On Cohen's view, liberal egalitarianism should agree with right-wing approaches about the need to align substantive responsibility and attributive responsibility, while additionally arguing that justice requires rectification in cases where inequalities are *not* due to choices for which individuals bear attributive responsibility.

The resulting family of views is sometimes described as responsibility catering egalitarianism, but much more frequently as luck egalitarianism—the term I shall use.[13] Two representative statements of this kind of position would be Temkin's 'what is objectionable is some being worse off than others through no fault of their own' (Temkin 1993: 17), and Cohen's claim that the purpose of egalitarianism 'is to eliminate involuntary disadvantage, by which I (stipulatively) mean disadvantage for which the sufferer cannot be held responsible, since it does not appropriately reflect choices that he has made or is making or would make' (Cohen 1989: 916). Thus, the luck egalitarian can agree that in cases such as the gardener and the tennis player, it would be fair to allow the inequalities to lie where they fall, but in cases where the inequalities are a matter of bad luck, then they would require correction as a matter of justice.

Blake and Risse (2008) make a helpful distinction between two different ways of thinking about justice: direct and indirect views. Luck egalitarianism is a direct

---

[13] Despite having inspired luck egalitarianism, Dworkin did not self-identify as a luck egalitarian, preferring to specify his views through the idea of fair insurance. For an overview of Dworkin's relationship to luck egalitarianism, see Arneson (2018).

view, while Rawls's is an indirect view. On direct views, claims about distributive justice can be derived directly from abstract moral principles, such as moral egalitarianism (the idea that individuals are all in an important sense equal), and the principle that it is bad if some are worse off than others through no fault of their own.

On indirect views, things are more complex. An account of the just distribution of a good cannot be derived merely from thinking about fundamental moral concepts. Rather, justice requires the maintenance of certain normatively relevant relationships among individuals. The presence of inequalities is not irrelevant to these relationships, but it is not directly and algorithmically relevant either. While a direct theorist might think that knowing only the distribution of a particular good would be sufficient to determine whether the distribution was fair, an indirect theorist would argue that justice is richer and more contextual. For the indirect theorist, justice in the distribution of goods will depend also on factors such as the scale of the inequalities, how they are interpreted by the individuals, and the implications that inequality in one sphere has implications for inequalities in other spheres.

The luck egalitarian response thus involved a double move away from Rawls's initial paradigm. First, it was a shift from an indirect to a direct account of justice. Second, it was a shift from taking an indirect account of justice to be fundamental and interpreting substantive responsibility contextually in the light of this, to taking as fundamental a pre-institutional view of responsibility that aligned substantive responsibility closely with attributive responsibility, and using that as an acceptability constraint on the construction of theories of justice. The underlying conceptions both of what a theory of justice is for, and what counts as a good reason for holding someone substantively responsible, differ significantly between these two approaches. For example, cases like the gardener and the tennis player presuppose rather than prove that substantive responsibility should be aligned with attributive responsibility. To the extent that it is better to think of substantive responsibility in institutional terms, thought experiments of this kind are likely to not only lack external validity but also beg the question.

This chapter does not aim to provide a full analysis of the merits of direct versus indirect accounts of justice. It focuses on the narrower question of how health systems should view substantive responsibility.[14] It argues that—at least in the context of public health—it is a source of weakness rather than strength if egalitarians closely align attributive and substantive responsibility.

Luck egalitarians are, as a fundamental component of their position, committed to the claim that whether a particular choice is attributable in the relevant sense makes a difference to whether any resulting inequality is fair or unfair. In framing

---

[14] Chapter 9 takes up the question of direct and indirect accounts of justice in its discussion of health-related inequalities, arguing that indirect approaches should be preferred.

attributive responsibility and justice-based rectification as mutually exclusive, luck egalitarian approaches thus need to walk a narrow path between the Scylla of judgemental harshness—where a policy ends up stigmatizing or otherwise penalizing people who were not sufficiently attributively responsible for their bad state—and the Charybdis of demeaning condescension—where a policy treats individuals as not capable or responsible, when they feel that they are.

One initial worry is that even within the simplified world of thought experiments, this approach may give the wrong results in the domain of health. Fleurbaey (1995: 40) introduces the case of Bert, who despite having had normal opportunities has 'freely adopted a negligent and reckless character'. Loving the feel of the wind in his hair, he deliberately rides his motorcycle without a helmet, while knowing that this is illegal. Bert has an accident that is caused by his carelessness, which results in a severe head injury. He cannot afford to pay for the surgery, and does not have health insurance. Given that Fleurbaey has stipulated in the setup of the case that Bert is as responsible as anyone reasonably could be for the bad that befalls him, it is initially hard to see how the luck egalitarian could avoid judging that Bert's injury arises through his own choice, and that putting it right would not be a matter of justice.

This conclusion seems to many to be rather harsh, and to be contrary to what justice requires. In response to this kind of case, luck egalitarians have sometimes argued that while there is no obligation of justice, there may be duties of charity, or of solidarity in such cases. Alternatively, other luck egalitarians have argued that we should be pluralists about the content of justice (Segall 2007), or have doubted the practical relevance of such cases by arguing that it is *choices* rather than each individual choice that luck egalitarians should aim to be sensitive to (Cappelen and Norheim 2005).

Avoiding demeaning condescension will also prove a challenge. Barry (2006) provides a radical defence of luck egalitarianism against the charge that it is overly individualistic and insufficiently attentive to structural injustice, by arguing that the nature of structural inequalities in the world entail that the conditions that a sensible luck egalitarianism would require for responsibility are in fact rarely present. On this view, individuals are rarely responsible for any of their choices: 'applying this theory leads to the conclusion that few of the inequalities that exist today are acceptable. The genius of luck egalitarianism is that it highlights the radical implications of any theory that is truly sensitive to individual choice' (Barry 2006: 102). If accepted, this argument would indeed very significantly enlarge the range of circumstances that are assigned to brute luck, and so would undercut applications of luck egalitarianism that would license a judgemental attitude towards those who end up badly off. However, it would nonetheless be a bad response to the central problem of how policymakers should think about substantive responsibility, and not only because it would be unlikely that Barry's approach to attributive responsibility would ever be adopted in practice.

The claim that society is so unjust that hardly anyone is ever attributively responsible for anything risks discounting and demeaning the agency that those in straitened circumstances do enjoy.

Strawson (1974) argues persuasively that the idea of responsibility as attributability is core to ideas of human dignity and of morality. While it may be unpleasant to be held responsible by being blamed, the possibility of praise and blame, and of being held accountable more generally, is an inextricable element of rationality and human dignity. For example, to treat someone as not responsible under the criminal law, or as lacking capacity to make decisions about their life, is by the same token to treat them as lacking full autonomy. Thus, denying that attributive responsibility has a significant role to play in human life risks undermining the sense of equal dignity that egalitarian justice seeks to defend. Resisting the rise of personal responsibility in public policy discourse by arguing that hardly anyone has attributive responsibility for their choices or outcomes would, at best, provide a Pyrrhic victory.[15]

Knowing how, and where, to draw the line between 'choice' and 'circumstance' would be crucial if luck egalitarianism were to be useful in helping policymakers to make wise decisions. Outside of the context of thought experiments in which it can be stipulated that a particular outcome is due either to choice or to circumstance, fairly categorizing outcomes is challenging. Luck is difficult to define, and even more difficult to measure. This implementation question has been largely ignored within the philosophical literature—though it has received some debate in the crossover between economics and philosophy.[16] The result is that there are a wide variety of potential policy positions that could be taken to be an expression of what luck egalitarianism requires, depending on the model of responsibility adopted, and the conditions under which attributive responsibility is assigned: 'at the level of real world outcomes, luck egalitarianism is a procedural gamble

---

[15] Darren McGarvey, in a memoir about growing up in poverty, abuse, and drug addiction in Glasgow, explains it as follows: 'It's counter-intuitive to accept responsibility for certain things, particularly when our circumstances are beyond our control. This is particularly true if we have suffered abuse, neglect or oppression. But striving to take responsibility is not about blame, it's about honestly trying to identify what pieces of the puzzle are within our capacity to deal with. This approach is far more radical than simply attributing responsibility for every ill in society to a "system" or vaguely defined power dynamic—something we lefties have gotten all too good at' (McGarvey 2018: 177).

[16] In one of the most influential accounts within the philosophical literature, Roemer argues that the fair thing to do is to hold someone substantively responsible for the degree to which they raise their risk level above a baseline set by other individuals in a similar position to them (Roemer 1993). One crucial, and unresolved, problem for any such approach is that each individual belongs to multiple reference classes—for example as a man, or a teacher, or someone aged 20–30, as working class, or as a male working-class teacher aged 27, or as a recreational drug user. The reference class a person is assigned to will make a significant difference to what the baseline risk rates are taken to be, and there is not a single uniquely correct answer to which reference class an individual belongs (Hájek 2007). In a more recent review article on the topic, Roemer and Trannoy (2016) conclude that the main focus of equality of opportunity should be on 'equalization of outcomes that are due to differential circumstances', and that working out when poor choices should not be compensated should receive rather less attention (Roemer and Trannoy 2016: 1328).

that can lead to anything from all-embracing welfare structures to very unforgiving and harsh models' (Ahola-Launonen 2018: 48).

Some have argued for a division of labour—claiming that this is not strictly speaking a problem for luck egalitarians, as 'luck egalitarianism is a view about *how to respond* to responsibility, not a view about who is in fact responsible for bringing about what' (Albertsen and Knight 2015: 167). In so far as luck egalitarians foreground the claim that distributions ought to be responsive to choice and responsibility, but either think it is someone else's task to define who is responsible for what, or interpret responsibility in such a disparate variety of ways that policymakers can choose the conception of responsibility that best coincides with their particular political views, there is a clear and present danger of their strategy backfiring and further undermining the position of the worst off. In particular, policymakers may in practice place more emphasis on individualistic models of responsibility, even if sophisticated luck egalitarians would wish to emphasize a richer and socially engaged model (Voigt 2013: 154). In circumstances of significant inequalities of power, the luck egalitarian approach may end up unwittingly further entrenching structural injustices.

In any case, it seems plausible to suppose that if state officials are to make accurate determinations of the degree of attributive responsibility that citizens have for adverse events that befall them, this would require access to a very significant amount of private information about each citizen. This is likely to be incompatible with the respect for privacy that the state is required to show its citizens (Wolff 1998; Anderson 1999). This is a particular concern in the domain of health, given the sensitivities around medical confidentiality and the doctor–patient relationship. Obviously, where a patient thinks that their doctor is going to be judging the responsibility of their choices, and potentially penalizing them for these, it is hardly likely to inculcate an open and trusting approach required for a good therapeutic alliance. It is thus unclear how information that allowed the degree of individual responsibility for particular episodes of ill-health to be measured could be gathered and used without undermining trust.

Given the difficulty of getting hold of the kind of information that would be required to make accurate decisions about where to draw the line between choice and circumstance without disproportionate interferences with privacy, policies that foreground personal responsibility may end up being punitive, arbitrary, or socially corrosive—in a word, irresponsible. Mounk (2017) notes how in public policy debates in the USA, it is often the case that poorer citizens will be blamed if *any* of the links of the causal chain that ended in them being worse off could be attributed to them. This is unreasonable, given that there will be a number of other factors that may have played an equally large or larger role in the adverse event coming about. Further, it is often groups who are already subject to stigmatization, such as drinkers, smokers, and the obese, who are singled out in debates about responsibility and health. These stigmatized groups represent just a few of the many of the ways in which individuals could be claimed to increase their health risks. So there are obvious worries about whether such selective targeting of alleged failures of personal responsibility will amount to scapegoating or blaming the victim (Friesen 2018).

8.6 Allowing, Rather than Forcing, People to be Responsible

We started this chapter with an overview of the variety of different practices of holding accountable—including practices of deliberately not holding individuals accountable even when they are attributively responsible and practices of holding accountable even without attributive responsibility—and explained how each of these different approaches could be seen as coherent and intelligible interpretations of responsibility, given the varied purposes of public policy. Against this contextualism, luck egalitarianism favours a top-down approach, in which attributive responsibility and substantive responsibility are closely aligned. One implication is that luck egalitarian approaches find it difficult to explain how, or even to acknowledge that, there can be areas of policy in which the interests of justice will be best served by deliberately *not* seeking to condition what the state owes to individuals on judgements of attributive responsibility.

Institutional accounts of substantive responsibility allow for a much greater degree of contextual flexibility. How closely attributive and substantive responsibility should be aligned in particular instances will depend on context-specific features such as the policy aims, the likelihood that individuals will act in a way that is conducive to the goals of the policy without being held accountable through judgements of substantive responsibility, the extent to which holding individuals substantively responsible would make them more likely to act in a way conducive to the policy's success, and the costs (both financial and ethical) of decision-makers obtaining the information that would be necessary to make sufficiently accurate judgements of attributive responsibility.

Criminal justice policy is one policy domain in which institutional accounts will agree with luck egalitarian accounts that there are strong reasons to keep attributive and substantive responsibility closely aligned. Aligning attributive and substantive responsibility by ensuring that all and only the guilty are punished is generally agreed to be a prerequisite for fairness in criminal justice. Given the kinds of unfair advantage that can be gained by lawbreaking, another main aim of the criminal justice system is provide incentives for compliance with the law—by bringing those responsible to justice, and ensuring that those who are found guilty are appropriately punished. So in the context of criminal justice, there are both intrinsic and instrumental reasons for aligning attributive and substantive responsibility.

Public health policy, in contrast, aims to protect and promote population health by improving the health of individuals in a way that respects individual liberty and autonomy. It is no part of the ethos either of medicine or of public health to think that all and only those who have not brought their disease on themselves should be helped. Indeed, even in cases such as Fleurbaey's Bert, where it is deliberately stipulated that he acts negligently, there are few who think that the right thing to do would be to withhold treatment. Moreover, the individuals whom public health aims to benefit already have a strong incentive to protect their own health: to the extent that they are significantly worsening their own health and well-being they already are suffering or are likely to suffer significant ill effects. If the fact that someone is shortening their own life, or increasing their own future pain and immobility does not provide sufficient intrinsic motivation, it is not very clear that holding them accountable for their failure to protect their own health would provide an effective additional incentive.

Overall, on an institutional account, there are good reasons for thinking that attributive and substantive responsibility should be closely connected in criminal justice, but also good reasons for thinking that they should not be closely aligned in health policy. This leads to a fundamental question: would it be better to replace the existing variety of different contextual understandings of responsibility with a single standard derived from luck egalitarianism (and thus jettison the institutional account of responsibility), or keep hold of the contextual accounts and jettison luck egalitarianism?

Answering this question definitively is beyond the scope of this chapter, but we will note some reasons for thinking that the right to public health account (and healthcare more broadly) do require substantive responsibility to be widely divorced from attributive responsibility. To the extent that these considerations are persuasive, one inference to draw would be that luck egalitarians misinterpreted the challenge posed by the rise of the discourse of personal responsibility. Perhaps the correct response would not have been to incorporate the idea of personal responsibility into egalitarian theory, but to re-articulate a convincing solidaristic vision of society in the context of rising inequality and greater cultural differentiation, and use this to explain and justify a variety of models of substantive responsibility.

On the right to public health account, the main aim of public health is to construct systems in such a way that, as far as is compatible with other ethical requirements, whether good health outcomes result for individuals is *not* contingent on particular choices that individuals make. Individuals should not be put in dangerous or unsafe situations to start with; and to the extent they are, they should have backup and others to rely on and help them to face these dangers. The challenge is how to engineer systems that, while preserving liberty and autonomy as much as possible, make it unlikely that given a range of expected human behaviours, anyone gets harmed or injured, and that even if they do they will be up and functioning again as soon as possible. This may on occasion make it more difficult for the person who doesn't want to be helped; but as argued in

Chapters 5.7 and 6.5, the fact that someone is annoyed by measures put in place to protect the health and safety of others need not mean that these measures are illegitimate. The right to public health is important; and ensuring the safety of each individual's life is a core duty of responsible societies.

Nowhere in this approach to designing safety and equal citizenship into systems need there be a denial of the importance of attributive responsibility. The commitment to safety and to picking people up when they fall is there precisely because of the importance of each individual being able to lead the kind of life that they have reason to value. It is entirely possible to take attributive responsibility seriously, while also being strongly committed to ensuring that acting 'responsibly' is not a necessary condition for having a good life or healthy life. In so doing, we can hang on to the elements that matter most about responsibility—individuals being able to alter in a positive manner the way that their lives go—without committing to the implausible view that no one is ever responsible for anything.

One way of drawing this distinction is between *allowing* people to be responsible, in that what they do and choose will make a meaningful difference to their lives, and *forcing* people to be responsible by making them navigate a narrow precipice alone, and from which any slip will be catastrophic. The right to public health account takes away the irresponsible feature of policies of responsibilization—the idea that a society is better the more weight can be piled on to the idea of individual responsibility—while retaining responsibility as an ethical virtue and ideal.[17]

Public health and healthcare more broadly should thus simply refuse the suggestion that the domain of valid claims of justice starts where attributive responsibility ends. Looking in a little more detail at chronic disease care allows us to see why. The way in which care is delivered for most chronic illnesses within modern healthcare systems would make little sense without the assumption that patients are able to assume attributive responsibility for adhering to treatment advice (Walker 2019: ch. 6). In diseases such as diabetes or HIV, the day-to-day control and management of the disease is largely left up to the patient. While the patient may see a specialist every few months, the main factor that will make a difference between the patient keeping on an even keel and deteriorating is how well they adhere to the treatment regime. For example, whether the diabetic is in reasonably good control of their blood sugar levels, or the HIV patient takes the medication on schedule to ensure that the virus does not develop resistance. Neither of these will reliably occur unless patients are not only able to take

---

[17] Dewey puts this point well: 'One is held responsible in order that he may *become* responsible, that is, responsive to the needs and claims of others, to the obligations implicit in his position. Those who hold others accountable for their conduct are themselves accountable for doing it in such a manner that this responsiveness develops. Otherwise they are themselves irresponsible in their own conduct' (Dewey and Tufts 1981: 304–5).

control of their treatment regime, but in fact successfully do so (at least enough of the time). So the idea that it would be sensible to make a blanket judgement that patients are not attributively responsible for adhering to care would appear to be a non-starter.

At the same time, all the evidence suggests that rates of adherence for long-term treatment are in fact fairly low (no more than 50 per cent in most cases), and that patients will often not be fully honest with healthcare practitioners about the extent of their adherence. An array of social factors will be linked to adherence rates—including features such as the side-effects of drugs, deep-seated views about the badness of being reliant on medication, religious views, and broader social determinants of health. Notwithstanding this evidence on adherence rates, few thoughtful commentators think that attempting to condition a health system's response on a determination of which patients are attributively responsible for their failure to adhere would be helpful for patients, or would allow healthcare professionals to do their jobs more effectively. Patients should be offered high-quality care and support, regardless of the extent of their adherence.[18]

8.7 Conclusion

Responsibility in one form or another is central to ethical reasoning. This chapter has argued strongly against the narrowing of the idea of responsibility that has happened since the early 1980s as societies have become more unequal. Discussion of responsibility within the political context has increasingly come to be seen as a matter of personal responsibility, where personal responsibility has been thought of mainly under an individualistic model that isolates out individuals in order to blame them. When applied to health, such approaches are highly likely to be unfair, in so far as they place the blame on the individual and screen out all the other parts of the complex systems in which they are embedded. They are also highly likely to be counterproductive.

Luck egalitarian approaches, which aim to combine a commitment to taking personal responsibility seriously with an egalitarian distribution outlook, will struggle to avoid both harshness and being demeaning, and there is no reason to think that they will provide an enlightening approach to responsibility in public health. It will be better to adopt an indirect approach to justice, and to think about substantive responsibility in institutional terms. The right to public health account will recommend an approach to substantive responsibility that allows, rather than forces, individuals to take responsibility for their health, though—as

---

[18] The kinds of support available will make a significant difference to patients' ability to adhere, and that the ability to adhere will itself have a social gradient (Stutzin Donoso 2018). This makes it more rather than less important to provide tailored support.

we shall see in Chapter 10—it will also keep some room for substantive responsibility at an individual level where this is necessary to ensure herd immunity. Its main focus for substantive responsibility will be on holding states and corporations accountable.

The combination of rising inequality, reduced solidarity, and more individualistic and judgemental models of substantive responsibility, which as we saw led to the decline of the social democratic model, itself creates a deep political challenge for public health. Rethinking health equity, and how best to pursue this goal within a broader institutional context that may not be particularly friendly to public health, is vital, as the next chapter examines.

9

Measuring and Combatting Health-Related Inequalities

9.1 Introduction

An indefinitely large range of factors, from the genetic makeup of individuals, to social structures and global policies, directly or indirectly affect each individual's health. These factors obviously include policies within healthcare and public health more broadly, but also include features such as the design of transport policy, the structure of the taxation system, housing policy, and employment law. Factors from within the health domain narrowly conceived will also affect variables outside of this domain such as income or transportation choices, and then in turn further shape health outcomes narrowly conceived.

As should be familiar by now from thinking in systems terms, the distinction between inputs and outputs is not an absolute one: what is an output for one part of a system will be an input for another. To give a very simple example, if a health system operates according to the inverse care law (Tudor-Hart 1971), then there will be a feedback loop between the number of doctors per capita in a locality and the health outcomes. Not only will the number of doctors be an input to the achieved health outcomes, but the health outcomes will be an input to the maintained number of doctors per capita. Feedbacks of similar kinds—though often more complex in their mechanisms—are pervasive. Drawing out all the factors that are relevant, and their interrelations is a huge task, as the effects of these factors will not function uniformly across different stages of the life-course, or across different ethnic groups or geographic locations.

One implication is that while particular types of inequalities in health, such as inequalities in life expectancy between poorer and richer parts of a city, are often singled out for discussion, there are in fact indefinitely many different inequalities that could be discussed. Even confining attention to healthcare systems in a narrow sense, variables commonly measured by health systems include life expectancy, healthy life expectancy, rates of maternal mortality, five-year survival for bowel cancer treatment, rates of surgical complications, numbers of medically induced bankruptcies, per capita health spending, number of doctors per capita, numbers of hospital beds per capita, percentage of vaccine coverage, mean levels of attendance at primary care, and levels of satisfaction with care. Each of these measures, and the thousands of others that are used within health systems and beyond, and the indefinitely greater number of variables that could be measured but currently are not, will reveal inequalities.

This chapter examines health-related inequalities in more detail. I use the term 'health-related inequalities' in a way that is deliberately inclusive and open-ended—as part of what needs to be analysed is how broadly it is useful to spread the net in thinking about justice, health, and inequality. How should health systems measure health-related inequalities, decide which health-related inequalities are unjust, and how to prioritize the reduction of unjust related inequalities? The sheer range of relevant factors, and of potential intervention points, makes the need for an ethical framework for thinking about health equity pressing.

As Chapter 8.5 explored, there are two main orientations within philosophy as to the meaning and implications of equality—direct and indirect accounts of justice. Direct views of justice take equality to be fundamentally a measure that applies to states of affairs, and to distributions of things. On such a view, things are fair when they are distributed according to an abstractly described pattern; if such an ideal pattern is assumed to be equality, then any inequalities in goods that are relevant to well-being are pro tanto objectionable. Luck egalitarianism is formed by combining a direct view of justice with the idea that the ideal pattern should be sensitive to personal choices.

Indirect views of justice argue that equality should be understood as a fundamentally social good. On such views, we will need to know rather more to understand whether, and if so how, a particular inequality is objectionable. On indirect views there can be a variety of reasons for objecting to inequality, in addition to the mere fact of inequality. These can be connected to how the inequalities come about, or their effect, or what the fact of the inequality expresses (Anderson 2010: 2–3). For example, inequalities in particular goods may make it likely that there will be a violation of equal concern, or gross inequalities in power and ability to shape democratic institutions, or objectionable inequalities in status (Scanlon 2018).

This chapter argues that what is needed for a viable policy-relevant account of health equity is an attention to structures and processes and their mutual interactions, and how states should intervene to improve individuals' lives. Direct approaches to justice are much less able to do so than indirect approaches. Direct approaches tend to make the questionable assumption that it is inequalities only in either a single or a small number of master variables that matter for their own sake. Direct approaches also frame questions of justice in a way that deliberately abstracts from structures and processes, and unfortunately do little if any of the work that would then be required to explain how the theory could be used to guide policy within complex systems that are shaped by performativity. Indirect approaches do not share either disadvantage. The chapter closes by sketching an indirect account of health justice, which is designed for processes, structures, and performativity, and takes health-related stigma as a case study.

9.2 The Concept of a Health Inequity

This book presupposes rather than seeks to justify the status of equality as an ethical ideal. Inquiry needs to start somewhere, and the idea of equality is now so widely accepted that as Amartya Sen put it, 'a theory of justice in the contemporary world would not have any serious plausibility if it did not value equality in some space—a space that would be seen as important in that theory' (Sen 2004: 22). Contemporary philosophy tends to focus on how equality should be interpreted as an ideal, rather than whether it is justifiable. Indeed, the ethical analysis provided in Part II of this book presupposed a background idea of equality both in justifying the right to public health, and in specifying how to prioritize health risks.[1]

Anglo-American political philosophy since the 1980s has tended to assume a particular interpretation of equality as a political ideal—namely one that conceives it through the lens of a direct theory of justice. Looked at from this perspective, the most fundamental question in political philosophy seemed to be, in Amartya Sen's words 'equality of what?' (Sen 1979). Within these debates, it was taken for granted that justice requires distributing *something* in such a way that individuals have the correct share of it. G. A. Cohen famously labelled this the question of the currency of egalitarian justice: '[w]hat aspects of a person's condition should count in a *fundamental* way for egalitarians, and not merely as a cause of or proxy for what we regard as fundamental?' (Cohen 1989: 906).[2]

On a direct view of justice, the most important question about justice and health would be the relationship between health (however conceived) and the currency or currencies of justice (Wilson 2009b, 2011b). Debates within epidemiology have started from a different place, namely distinguishing health inequalities from health inequities. Health inequality is used as a normatively neutral term to 'designate differences, variations, and disparities in the health achievements of individuals and groups' (Kawachi et al. 2002: 647). Health inequities are a subset of health inequalities, namely those which are, all things considered, unjust. However, in identifying health inequities epidemiologists have not usually felt the need to do so by reference to a currency or currencies of justice.

It is agreed—both in the epidemiological literature and in the philosophical literature—that some degree of health-related inequality is the unavoidable corollary of diversity. Not every health-related variable could even in principle be equalized at the same time. Even assuming that—impossibly—a decision-maker were to have a total and certain understanding of causal relations within health

---

[1] For discussion of the nature and justification of equality as an ideal, see Wilson (2007c, 2007d); O'Neill (2008); Waldron (2017).

[2] In formulating his view, Cohen (1989: 906) explicitly states that 'I take for granted that there is something which justice requires people to have equal amounts of.'

systems, it would not be possible to equalize all health-related measures simultaneously, due to variable rates of conversion between different factors.[3] To take a highly simplified example, suppose that a decision-maker cared only about two kinds of inequalities: inequalities in healthcare spending on individual patients, and inequalities in patients' health outcomes. Since some treatments are significantly more expensive than others, and some individuals will also get ill more often than others, it would be simply impossible simultaneously to equalize both the spend per patient, and also the health outcomes of individual patients. In real-world scenarios, equalizing all health-related inequalities would be even more obviously unfeasible.

In making sense of how states should respond to the inevitable fact that making some factors more equal will make others less equal, it is useful to distinguish between treating individuals as equals, and treating every individual in the same way. For example, it is widely assumed to be compatible with treating individuals as equals (or even required by it) to give a state pension to those who have reached a certain age threshold but not other citizens; to give additional educational support to those with special needs; to allow to vote all and only those who are above the threshold age and not otherwise disqualified; and to punish those who break the law but not law-abiding citizens.

In explaining why this is, Dworkin (1977: 227) makes a distinction between the right to *equal treatment*, conceived as 'the right to an equal distribution of some opportunity or resource or burden' and the right to *treatment as an equal*, conceived as 'the right, not to receive the same distribution of some burden or benefit, but to be treated with the same respect and concern as anyone else.' Dworkin argues that it is the right to treatment as an equal which is more fundamental, and that this will sometimes involve treating individuals differently. He motivates this thought as follows: 'If I have two children, and one is dying from a disease that is making the other uncomfortable, I do not show equal concern if I flip a coin to decide which should have the remaining dose of a drug. This example shows that the right to treatment as an equal is fundamental, and the right to equal treatment, derivative' (Dworkin 1977: 227). This distinction between treatment as an equal, and equal treatment, was implicit in the discussion of features such as severity and relative capacity to benefit in Chapter 7.

Some goods are also much more easily distributable (and redistributable) than others. If a government wanted to greatly reduce inequalities in income, it could do so by redistribution—for example, by creating a progressive income tax system, and distributing the resulting money to the poor through the benefits

---

[3] As Jonathan Wolff points out, the underlying observation in one sense goes back to Marx's Critique of the Gotha Programme, but while Marx took it to show a fundamental problem with the idea of making people equal, in the 1980s analytic political philosophers such as Sen, Dworkin, Cohen, and followers 'took it as the start of a research program to capture the true nature of egalitarianism' (Wolff 2015: 210).

system. Healthy life expectancy is not directly transferable from one person to another, except in rare cases such as organ transplantation, and so it would be much more difficult for a government to greatly reduce such inequalities. Moreover, some diseases are incurable and lead to a very significantly shortened life—and so make it impossible to achieve equality of outcome even of one variable such as life expectancy, unless everyone else is made very significantly worse off. Short of eliminating genetic diversity through a policy of mass cloning, and elimination of socially determined variation by a relentless programme of total standardization, a significant degree of variation in health outcomes seems unavoidable (Whitehead 1990: 6–7).

The combination of differential conversion factors, and the non-redistributability of important health goods, implies that it would not be a desirable or even a coherent project, to eliminate *all* health-related inequalities. This argument establishes that there will be some health inequalities that are not health inequities, but it does not give a positive criterion by which to identify health inequities, or give insights into which health-related inequalities should be allowed to persist in order to reduce or eliminate health-related inequities.

In a very widely cited article in the epidemiological literature, Whitehead argues that health inequities should be defined as differences in health that are 'unnecessary and avoidable but, in addition are also considered unfair and unjust' (Whitehead 1990: 5).[4] On this view, there are two main ways in which health inequalities can fail to be health inequities. First, if the inequality is either necessary or unavoidable, and second if the inequality is not unfair. Whitehead cites skiing injuries—at least where the activity is undertaken by the wealthy for leisure purposes—as being paradigmatic of cases where the injuries suffered more frequently by some groups rather than others is not unfair, since skiing is 'widely viewed as a voluntary activity chosen by those who accept and insure against the risks involved' (Whitehead 1990: 220). This threatens to make the kind of confusion between attributive and substantive responsibility that was criticized in Chapter 8. While the claim that an activity is voluntary may be relevant to judgements of substantive responsibility, there is no necessity to regard it as a decisive consideration.

Whitehead also draws some strong, and rather implausible, conclusions about the nature of health inequities from the requirement that the inequalities be 'unnecessary and avoidable', including that the 'portion of the health differential attributable to natural biological variation can be considered inevitable rather than inequitable' (Whitehead 1990: 6–7), and that it is only where 'social

---

[4] It is unclear why the definition refers to differences which are both unfair and unjust, given that Whitehead does not make any distinction between these two terms, which are in any case close to synonymous in ordinary usage. I shall assume that we should delete unfair, and simply leave ourselves with the idea of unjust differences.

processes...produce health differences rather than these being determined biologically' (Whitehead and Dahlgren 2006: 2), that it makes sense to say that there is a health inequity.

I have criticized these mistakes in detail elsewhere (see Wilson 2011b; see also Smith 2015). For present purposes it is important to point out that the fact that something is caused by 'nature' it follows neither that we are powerless to stop it, nor that we should refrain from trying to stop it. Indeed, all of medicine could perhaps be thought of as the attempt to stave off what would otherwise be the inevitable operation of nature (Mill 1985 [1874]). Moreover, even if it would be unfair and undesirable to attempt to equalize *all* health inequalities, it would not follow that those who—as a result of the genetic lottery—are born with a debilitating and incurable health condition have no claim to some form of rectification on grounds of justice.

Even if policymakers are not in a position to cure the incurable, there are myriad other things that can be done to ensure that physical and mental impairments do not prevent individuals from functioning as equal citizens. Inequalities in health and quality of life as a result of disability have increasingly been recognized as a potential violation of human rights—especially since the ratification of the UN Convention on the Rights of Persons with Disabilities (2006). Indeed, as Chapter 8 discussed, luck egalitarians explicitly claim that disadvantages due to bad luck do raise issues of justice.[5] The claim that luck egalitarianism's central claim is false, and that in fact it is only inequalities with a social cause which are of concern for health equity seems unnecessarily controversial for an account of the *concept* of a health inequity. It would seem to be a mistake to build a controversial claim about the nature of egalitarian justice into the concept of a health inequity.[6]

Thus, a definition of health inequity is required that is both more ecumenical and more precise than Whitehead's. It is better to say that a health-related inequality is a health inequity if and only if it is an inequality that a just society would seek to counteract (Wilson 2011b). Working out which health-related inequalities a just society would seek to counteract requires not just advancing an account of justice, but also advancing an account of which health-related variables to measure and how to do so, and which kinds of interventions to favour. As Chapter 8 argued, adopting a strategy that places most weight on primary prevention at a

---

[5] See, for example, Dworkin (1981a, 1981b); Arneson (1989); Cohen (1989) for seminal contributions to the luck egalitarian literature, and for critiques of this position, see Anderson (1999); Scheffler (2003). Segall (2009) provides a useful examination of luck egalitarianism in the specific context of health.

[6] I argued against luck egalitarianism as a way of incorporating judgements of substantive responsibility into public health policy in Chapter 8, and in this chapter will argue against direct approaches to equality in health policy. Nonetheless, I think it is a mistake to define health inequity in a way that rules out on purely conceptual grounds some things that luck egalitarians think would be health inequities. The problem with luck egalitarian approaches to health policy is not that they are conceptually incoherent.

population level by systematically reducing risks, rather than placing such responsibility on the shoulders of individuals, is likely to be superior in terms of equity, cost-effectiveness and reduction of the individual burdensomeness of public health measures. Of course, even with such an account of justice, policy-makers would not be able to counteract all such inequalities simultaneously. An account of prioritization will also be required, which could draw on the account of which risks to health are most important to eliminate that was put forward in Chapter 7. The relationship between different risk factors that may combine to magnify risk is also relevant, as Section 9.5 discusses, and is taken up in the context of communicable disease in Chapter 10.

9.3 Direct Views of Justice, and their Implications for the Study of Health Equity

Should health—however health is specified—be taken to be a currency of justice? Making the case for health as a currency of justice would first of all require clarifying which aspect or aspects of health 'should count in a *fundamental* way for egalitarians, and not merely as a cause of or proxy for what we regard as fundamental' (Cohen 1989: 906). The following section considers some difficulties in specifying what such a measure would amount to. Even without going into these specifics, it should be apparent that it is implausible to think that health is the *only* currency of egalitarian justice. Regardless of how health is conceived, there seem to be goods that are important to a just society which are neither reducible to fair distribution of health, nor valued only for their contribution to fair distribution of health. For example, it would seem strange to describe a society which was rife with racism and discrimination, and prevented women from voting or from holding political office, but yet where fortuitously everyone had the same level of health achievement, as one which was just in an egalitarian sense. Distributive egalitarian justice must care about more than merely health.[7]

The chief challenge for any direct approach to justice in health will be to explain how the normative importance of health relates to that of other goods that are important from the perspective of justice. There are two major options while sticking within the 'equality of what' framework. The first option would be that health (however defined) is not a currency of egalitarian justice. On such a view, health would be important from the perspective of justice only for the

---

[7] It has sometimes been suggested that on the WHO definition, namely that health is 'is a state of complete physical, mental and social well-being and not merely the absence of disease or infirmity' (World Health Organization 1948), health could plausibly be the sole currency of justice. If one were tempted by the idea that there should be a single currency of justice, then well-being, or some specification of it such as equal opportunity of welfare (Arneson 1989), might be a plausible way of so doing. However, as Chapter 1.2 explored, whether (re)defining health in such a way that it becomes coextensive with well-being is a good strategy is much less clear (Callahan 1973; Bok 2004).

impact that it has on the distribution of whatever good or goods do comprise the currency of justice. The second option would be that there are multiple (and mutually irreducible) currencies of egalitarian justice, and health (however defined) is but one of them. I shall briefly consider an influential version of each option for illustrative purposes, before arguing that the systemic nature of the social determinants of health makes it much more plausible to adopt an indirect perspective on justice.

Ronald Dworkin provides a good example of the first option. Dworkin picks out resources as the currency of egalitarian justice, using the concept of resources in a broad way to include not just 'external' resources such as land and money, but also talents, which are theorized as 'internal' resources. Disability and ill health are understood as negative internal resources (Dworkin 1981a, 1981b). Dworkin's view has two main implications for the study of health inequities. First, the duty of justice would be to ensure a fair share of resources for each person. Health would be one such resource, which could reasonably be traded against other goods. So the fair distribution of resources would not require policymakers to take health to be a special case, and to seek to equalize it separately from other goods.[8] Second, on Dworkin's view, the distinction between the natural and social is not of normative significance, and so should not be presupposed in a theory of justice. What matters according to Dworkin is whether someone can fairly be held responsible for the shortfall in their combined bundle of internal and external resources. If a person suffers a shortfall for which they cannot be held responsible, then this requires rectification, regardless of whether the cause of the shortfall is natural or social.

From the perspective of the right to public health account, focusing on equalizing resources in the way that Dworkin suggests looks myopic: it would appear to overlook the ways in which structural determinants of health, such as racism or mental health stigma, affect how and whether a given individual is able to convert resources into well-being. Moreover, in cases where racism is a structural determinant of ill health, it would seem mistaken to think that an adequate solution would be for resources to be given individually to affected individuals in order to compensate them for the 'negative resource' of being subject to discrimination. The conception of responsibility that lies of the heart of Dworkin's account is also problematic in the case of public health: as Chapter 8 discussed, it is difficult to operationalize without its becoming moralistic and irresponsible.

Capabilities approaches, of which Amartya Sen (1999) and Martha Nussbaum (2000, 2011) are the preeminent exponents, provide a good example of the

---

[8] Dworkin denies that ill health and disability give someone a *direct* claim to equalization of resources, on the ground that this would amount to a 'slavery of the talented' (Dworkin 1981b: 312). He argues that justice requires only that each person be given sufficient resources to enable them to purchase insurance against ill health and disability. However, it is questionable whether Dworkin's argument for this claim is consistent with his broader position, as Cohen Christofidis (2004) argues.

second option. Capabilities theorists argue that the capability to live a healthy life of a normal length is one of several currencies of justice. They differ in two fundamental ways from Dworkin. First, they argue that resources are the wrong distributive space to be working in. For the capabilities theorist, what is of value to people is being able to be or do certain things, such as using their practical reason, or playing, or living a healthy life of an ordinary length, rather than simply having resources at their disposal. Second, capabilities theorists argue that we should be pluralists when it comes to justice: there are a plurality of capabilities which are jointly necessary for a flourishing human life, and we cannot fully compensate a shortfall in one capability with a superfluity of another. Given this broad pluralistic framework, the argument for why health is a functioning which matters fundamentally for egalitarian justice is simple:

> health is among the most important conditions of human life and critically significant constituent of human capabilities which we have reason to value. Any conception of social justice that accepts the need for a fair distribution as well as efficient formation of human capabilities cannot ignore the role of health in human life and the opportunities that persons respectively have to achieve good health—free from escapable illness, avoidable afflictions and premature morality.
> (Sen 2004: 23)

The capabilities theorists' understanding of the role of health resonates much more strongly with the right to public health approach, though it is worth noting a couple of differences of emphasis and interpretation. Capabilities theorists make a distinction between capabilities and functionings—arguing that justice requires providing each person with the *capability* to function in the valued way, not ensure that they do actually function in this way. They argue it would be a mistake for a society to attempt to ensure that everyone has the given functioning, as someone may legitimately choose not to exercise it. (For example, someone may wish to fast for religious reasons, and it would be wrong for a society to force the fasting person to have the functioning of being well nourished.) The right to public health approach focuses on ensuring health-related functionings—while adequately respecting liberty and autonomy—rather than on providing capabilities to be healthy. This is because, as Chapter 5 explored, the freedom to choose health risks does not seem to be valuable for its own sake.

Capabilities theorists disagree about whether there should be a fixed list of central capabilities. Nussbaum argues that there is a set of ten central capabilities that are the same the world over, and that justice requires that governments ought to ensure at least a decent minimum of each capability to every citizen, while Sen is much more inclined to allow democratic communities to make up their own minds about which capabilities are most important for them, and how to measure these. The discussions of constructivism and measurement in Chapter 7, and earlier in this chapter, should provide some reasons for thinking Sen's interpretation is more plausible.[9]

Will a direct model of justice—whatever the account of the currency of egalitarian justice it favours—provide a lens that will be helpful for policymakers? Direct approaches assume that there is something that counts as getting the distribution right (a particular pattern of distribution of the favoured good(s)), and that the core philosophical question is to specify what this pattern is in a way that avoids all counterexamples. One important difficulty, as Nozick (1974) was the first to point out, is that regardless of what the correct pattern is, the actions of individuals will tend to take the distribution away from whatever pattern was thought to be just, leading perhaps to significant concentrations of power and money (as in his famous Wilt Chamberlain thought experiment). A patterned distribution is not something that can be set once and for all; it will need to be actively maintained. This will clearly involve gathering a significant amount of information, but also redistributing whatever it is the theory says matters (or downstream determinants of that thing) from time to time in the light of the desired pattern.

Those who are not committed libertarians will find the conclusion that Nozick drew rather implausible—namely, the infractions of liberty required to maintain a patterned distribution over time would violate rights, and so patterned theories of justice should be abandoned in favour of a historical account of justice (which has no concern for the pattern of holdings).[10] But the underlying point—namely, that it is vital to consider not just what would be ideal, but also how to get from here to there, and to maintain things in the desired state once it is achieved—is core to the systems theoretic analysis that we have pursued throughout this book. Once what would be required to realize and maintain a particular pattern of distribution is lucidly understood, it may come to seem markedly less attractive, as was discussed in the case of luck egalitarian conceptions of responsibility in Chapter 8.

9.4 Measuring Health-Related Inequalities

Talk of health-related inequalities might initially suggest a binary interpretation: the distribution of a variable of interest is either equal or it is not. But while there

---

[9] Sen's more recent work (such as Sen 2011) is probably best interpreted as using a capabilities approach within the context of an indirect account of justice. Robeyns (2017) provides an excellent overview of different ways of constructing and specifying capability approaches. Integration of capabilities approaches with complex systems approaches are at a fairly early stage, but see Craven (2017) for a helpful attempt to do so. For a useful analysis of health as a capability, see Venkatapuram (2011).

[10] Nozick argues that so long as goods and resources have been legitimately acquired, and are then transferred justly, there are no further questions of justice in the patterns of distribution to be answered. For a classic analysis and critique of this argument, see Cohen (1995).

is only one way for a distribution of a particular good to be equal (everyone has the same), there are a myriad of different ways in which it can be unequal. So the number of possible equal distributions is massively outnumbered by the number of unequal ones.

It is, of course, very common to say that some distributions are more unequal than others, but as Temkin (1993) examines, it can be controversial which of two distributions is more unequal even in a thought experiment in which it is stipulated that all relevant facts are known. Temkin asks us to imagine a sequence of 999 worlds, each containing exactly 1,000 people. In world 1, there is one person who is badly off, and 999 who are well off; in world 2 there are two people who are badly off, and 998 who are well off; in world 3, there are three people who are badly off, and 997 who are well off... and so we continue until we get to world 500, in which 500 people are badly off, and 500 who are well off... and then we continue to world 998, in which 998 people are badly off, and only two are well off, and then world 999, in which 999 are badly off, and only one person is well off.

Clearly, the world in which 999 people are well off, and only one person is badly off will be the world that has the highest average welfare. But Temkin asks us to attend to a different and separate question: which of the worlds is the most unequal? It is important for Temkin's thought experiment that, for each of the worlds 'the members of that world are equally skilled, hardworking, morally worthy and so forth, and that the better off are not responsible for the plight of the worse off' (Temkin 1993: 27). So, in thinking about this sequence, we are supposed to attend only to the equality of the distribution of welfare, rather than to reflect on considerations about whether any of the individuals deserve either their good or their bad position.

Temkin shows how it could plausibly be argued (among other interpretations) that things get less equal as we go through the sequence; that things get more equal; and that they stay at the same level of inequality. One upshot is that even when thinking about the inequality in distribution of a single good between those who do well and those who do less well, we need to be able to account for and to explain the effects of at least two separate features: the extent of the variation between the top and the bottom, and the numbers of individuals who occupy these places.

When these differences in interpretation of which outcomes are most unequal are combined also with the sheer variety of different health-related variables that it might be useful to measure, 'it is questionable whether there is any single fact which reports how unequal health is in some population' (Hausman 2007: 47). Researchers and policymakers can construct validated measures of inequality for particular purposes, but none of these will be the *only* legitimate way of measuring health inequality.

Moreover, even where there are measures of health that are in common usage, such as the EQ-5D or the SF36, which measure health-related quality of life and are often used as the basis for resource allocation decisions, it is disputed exactly what we should take them to be measuring—for example, whether they should be taken to be measures of health itself or rather as measures of the *value* that individuals place on particular health states. Hausman (2015) argues plausibly that it is unclear what it is for one state to contain more health than another—except when looking at a very gross level. Because of this, he suggests that it is more likely that when individuals are asked which of two states contain more health, they end up stating preferences with regard to different health states—states, moreover, with which they may or may not have had direct experience. The upshot, Hausman argues, is that health policymakers should acknowledge that health economic instruments such as the EQ-5D measure judgements about the value of health states, rather than measuring health itself. If so, the key question for public policy is how the value of health should be measured, rather than how health itself should be measured.

This process of measurement will obviously need to be interpreted in a constructivist rather than a realist manner (on this distinction, see the discussion in Chapter 7.9).[11] Various features will influence design and choice of measure, including the purpose for which the measure of inequality will be used, the ease or difficulty of obtaining data of a sufficient quality, the health-related correlations that the measure-designer is interested in, and whether there is a need to generate statistics in a standardized format.

The debate about whether to measure univariate or bivariate inequalities in health provides a useful example. Univariate measures of health inequalities are measures purely of a single health-related variable across a population, without reference to any other variable. Income inequality is often measured in this way, via the Gini coefficient, which measures inequality on a scale from 0 (where there would be complete equality) to 1 (where a single individual would earn all the income). The coefficient is defined in terms of the relative mean absolute difference.

Bivariate measures of inequality combine a measure of health with another variable that is of interest: for example, examining inequalities in life expectancy by social class or by race. Most literature on health inequalities uses bivariate measures. Two particular focuses of attention have been bivariate measures of mean GDP per capita and life expectancy, and bivariate measures of socioeconomic

---

[11] Hausman argues that existing instruments measure the private value of health—that is, health's 'contribution to whatever the individual cares about or should care about' (Hausman 2015: 158), whereas what should be measured for resource allocation purposes is the public value of health, which is the value health should be accorded from the perspective of the liberal state. Hausman argues that the public value of health should be measured by the extent to which (a) suffering and (b) activity limitations are relieved. I argue in Wilson (2016) that while he may be correct to think that it is the public value of health that the state should care about, there is no reason to think that the public value of health should be constrained in the way he suggests. In one way, much of this book could be seen as an answer to the question of how to determine the public value of health.

status and life expectancy. These measures have been used to derive important high-level conclusions about development, and about the role of socioeconomic status in life expectancy.

The literature on GDP and health has shown a strong correlation between GDP per capita of a state, and life expectancy at the lowest levels of development, but a progressively less strong correlation between GDP per capita and health outcomes as states tend towards greater wealth (Deaton 2003). Absolutely low GDP per capita understandably entails a much reduced ability to build and sustain a functional public health system, and maintain the conditions in which the broader social determinants of health are met, as well as to the ability to fund a healthcare system in the narrow sense. But as GDP per capita rises through the World Bank upper middle-income level towards the high-income level, less and less of the variance in life expectancy can be explained via increased GDP per capita. For example, according to World Bank data, in the period 2000–17, both Costa Rica and Cuba had slightly better life expectancies at birth than the USA, despite having a GDP per capita of five times less, and seven times less, respectively (World Bank 2019b), facts that caused some soul-searching when the US Institute of Medicine and National Research Council (2013) benchmarked the USA's health performance against a range of other countries.

The literature on the socioeconomic status has shown that, despite the lack of correlation between GDP per capita and life expectancy *between* countries, differences in income *within* countries are strongly correlated with health outcomes. As a person's socioeconomic status increases, so, on average, her life expectancy improves and so also do a range of other important health indicators.[12] There is evidence of a socioeconomic gradient in health in all countries for which statistics are available—a gradient that is by no means confined to groups who suffer absolute deprivation, but also occurs within groups who are, in absolute terms, fairly well off (Marmot 2005). For example, the famous Whitehall studies showed a gradient in health among civil servants, all of whom were, in absolute terms, comfortably off (Marmot et al. 1978).

Both univariate and bivariate measures of health have limitations. Univariate inequalities in health do not by themselves tell researchers anything about the causes of variation. For example, they do not allow researchers to separate what is caused by social factors from what is caused by natural factors, nor to separate

[12] The term socioeconomic status is usually used to refer to 'the relative position of a family or individual on a hierarchical social structure, based on their access to or control over wealth, prestige and power' (Mueller and Parcel 1981). When defined in this way, socioeconomic status is difficult to measure accurately and so studies tend to use something easier to measure as a proxy for socioeconomic status. Popular candidates for proxy measures include level of education, current income, overall wealth, type of occupation, or some combination of these measures (Shavers 2007). Each of these measures has their advantages and disadvantages, and clearly the fact that different studies use different measures creates a degree of difficulty in comparing the results of different studies examining the relationship between socioeconomic status and health (Braveman et al. 2005).

'unnecessary and avoidable' inequalities from necessary and unavoidable ones (Asada and Hedemann 2002). So, univariate measures of inequality will be much more useful on some understandings of health equity than others. If, as was argued in Section 9.3, the justice of a society depends on factors far beyond the distribution of health, then univariate measures of health will by themselves provide very little information about what would be required to make a society more just (Hausman et al. 2002; Hausman 2007).

Prioritizing bivariate measures would make sense if the health inequalities that governments should aim to reduce are inequalities that arise in the intersection of two variables—such as life expectancy and race. Clearly, many are concerned if groups such as African Americans do worse than groups such as white Americans— particularly if there is reason to think that the groups who do worse in health terms also do worse on other measures outside of health. And so such a group-based bivariate approach has an initial plausibility. Moreover, bivariate measures give at least some indication of potential interventions to reduce health-related inequalities.

However, if researchers looked only at bivariate measures of health inequalities between groups, they would overlook many of the health inequalities that are present in the population. Focusing on mean differences between groups flattens out variation within groups. Thus, failing to attend to the inequalities within groups will 'mask part of the inequality present in the population' (Murray et al. 1999: 537). This is an important limitation, as there may be significant variation within groups—some subgroups or individuals will do either much better or much worse than the mean. These differences may point to important causal factors that had previously been overlooked.[13]

Bivariate measures of health inequality may also be normatively uninformative. If the inequalities measured by bivariate health inequalities do not matter directly from the perspective of justice, then it would be questionable to assume that justice would be better served by attempting to equalize inequalities in health as opposed to inequalities in whatever did matter for its own sake for justice (Hausman 2007; Wilson 2009b; Preda 2018).

It is helpful at this point to return to the fundamentals of systems thinking, and to remind ourselves that univariate and bivariate measures do not exhaust the space of possibilities. Systems thinking alerts us to the fact that researchers should expect the relationships between the variables that mediate health outcomes to be multifarious and to involve feedback loops, just as in other cases of complex systems. As Chapter 4 argued, even where there is knowledge about the effects that a shift in a particular variable has on a particular health outcome in one context, it

---

[13] The worry about variation within groups could be answered in part by performing sub-group analyses—but the fundamental point remains that the mean health achievement of a (sub-)group of which someone is a member will not always be a good guide to how particular individuals within the (sub-)group are faring.

should not simply be assumed that shifts in that same variable will play the same causal role in other contexts.

The rise of big data analytics has made it increasingly obvious that it is a mistake to think of studying individual differences as in competition with understanding group-level differences. As the scope, richness, and interlinkage of datasets has improved significantly, it has become clear that explaining health variance at an individual level will typically require a model that is *high dimensional* (that is, one which takes into account variation among a large number of different variables simultaneously), rather than low dimensional (univariate or bivariate). Predictions about the health outcomes of individuals may be better based on isolating near-neighbours of the individual, who within the parameter space set out by the high-dimensional model are located close to the individual of interest, than by attempting to deduce top-down on the basis of membership of particular group. In so far as researchers are interested in questions about the dynamics and causation of different health-related inequalities, then the more variables than can be accurately and combined into a model, the better.

In the light of similar concerns, Asada (2013) argues for a three-stage process to the measurement of health equity. Stage one would be the measurement of the desired health variables at an individual level, and an application of a regression analysis that explains the variations in the health variable of interest to the greatest extent possible. Stage two would be to classify the factors revealed by the regression analysis into ethically justifiable and ethically unjustifiable factors. Stage three would be to 'estimate unfair health for each person by removing the influence of the legitimate factors on health' (Asada 2013: 47). It is unfair health inequalities thus considered that should be the target for social remediation, according to Asada. Such a rich and high-dimensional dataset would also allow the measurement of bivariate health inequities—for example, determining how much of the measured unfair health inequalities at an individual level is due to race or to social class.

9.5 Structural Justice

It is time to turn in a more focused way to the relationship between direct and indirect accounts of justice. Philosophical debate amongst those committed to the direct approach has tended to presuppose that it is the question of the currency of justice that is fundamental, and has given much less emphasis to what policymakers should measure (and how) in order to provide the information necessary to take action to reduce unjust inequalities, let alone what interventions would be most effective in reducing unjust inequalities.

As we remarked in considering the practical implications of direct approaches to thinking about responsibility in Chapter 8.5 and about measuring claims in

Chapter 7.9, this gives the erroneous implication that the most important task is deriving an account of the correct pattern, and that how to specify, measure, and use the pattern to improve people's lives is someone else's task. This tends to underestimate the complexity and the trade-offs involved in measuring what matters, the difficulty of achieving policy change, and to overestimate the usefulness of idealized patterns.

Policymakers need to decide what to do given how things currently are. This will rarely require a fully articulated account of how the currency or currencies of justice would be distributed in an ideal world (Sen 2006, 2011). They require an ability to spot manifest injustices, and an understanding of what would count as an improvement, rather than an elaborate ideal theory fit for all possible worlds. Moreover, interventions within complex systems need to be iterative rather than one-off. Any changes policymakers enact this year are highly unlikely to act as silver bullets that will solve problems at a stroke.

In the face of this reality of policymaking as wrangling interacting processes that are only partly under political control, a central problem for direct approaches to justice is that they tend to view redistribution as an episodic activity, which is applied to static snapshots. As we have seen, there are indefinitely many variables that can be measured, which could be described as health-related. These variables can be measured either on their own, or in bivariate or higher dimensional ways. In the face of this bewildering array of possible ways of measuring health-related inequalities, and the complexity and variability of the causal mechanisms involved, it will be vital for any direct approach to justice that aims to be useful for policymakers to clarify the relationship between the high-level pattern in the currency or currencies of justice that would constitute idealized justice, and the lived realities of individuals' lives as measured by the more specific health-related variables.

There are two strong reasons for thinking that indirect approaches will be better suited to the problems policymakers need to solve, rather than direct approaches. First, as the discussion in Chapter 4 argued, performativity raises some deep challenges for direct approaches to justice if and where the shape and boundaries of ethical concepts are socially constructed and socially constituted rather than discovered. In such cases, the correct distributive pattern cannot be specified (and does not exist) separately from the particular self-understandings present in a context. We mentioned concepts such as trust, privacy, and reasonable expectations in this regard, and shall shortly argue that experiences of stigma and hierarchy are important causes of health-related inequalities that cannot adequately be theorized from within the confines of a direct approach.

Second, taking a complex systems approach makes it much more plausible to think that institutional structure and processes rather than abstract distributive patterns should be the main subject of justice. Getting these structures and processes right is the only plausible way of ensuring sustained and systematic improvements for those whom systems are currently failing.[14] While the pattern of a distribution of health or other goods may be evidence for claims about injustice, justice does not consist in a static pattern in the distribution of goods. From this perspective, starting from questions about the currency of egalitarian justice, and about which static distribution of health will instantiate the right patterns, is to be engaged in answering the wrong questions. What is much more important is an understanding of the structural drivers of inequality, and particularly where and how these can be amenable to change.

9.6 Stigma

Policymakers have significant power to reshape systems over time, but there are also important limits to this power. Where systemic features combine to form what Powers and Faden (2008: 76) describe as 'densely woven patterns of disadvantage', then health-related inequalities may be stubbornly difficult to reduce.

Achieving change in cases of significant health-related inequalities will thus often be much less like applying a force to get a cart moving down a well-maintained road, and more like rolling a stone up a steep hill. Even this image somewhat underestimates the difficulty of the task. The slopes and difficult terrain which challenge governmental attempts to dismantle systemic inequalities were not created by slow geological processes that predate human beings; they are created by us, and in certain respects they *are* us.

This has two implications. First, and most obvious, a theory of just distribution as provided by direct accounts of justice is likely to be of limited value unless supplemented by and integrated with an account of how to adapt it to the shape of the terrain. Second, human cultural constructions do not merely play an empirical role in constraining what can be done to reduce health inequalities, but may also constitute ongoing threats to health.

Stigma provides a useful case study of the sheer difficulty of correcting determinants of health that operate at a socially constituted and structural level, both because of the multifarious ways in which stigma is intertwined with poor health and the ways in which it both relies on and further entrenches power imbalances. Much of the modern debate on stigma starts from Goffman's (1963) seminal work. Goffman argued that to be stigmatized is to be treated by others as 'of a less

---

[14]   As Young puts it: 'Ultimately, the judgements of justice, then, are not about the distributive patterns. Each distributive pattern only offers a piece of the puzzle, a clue to an account of generalized social *processes* which restrict the opportunities of some people to develop their capacities or access benefits while they enhance those of others…the purpose of equality is less to identify unlucky sources of inequality than to identify how institutions and social relationships differentially conspire to restrict the opportunities of some people to develop and exercise their capacities and enact their goals' (Young 2001: 16).

desirable kind', and potentially 'quite thoroughly bad, or dangerous, or weak' in virtue of a discrediting attribute (Goffman 1963: 2). The stigmatized person is thus 'reduced in our minds from a whole and usual person to a tainted, discounted person' (Goffman 1963: 2–3). Stigma, Goffman argued, arises in the context of relationships in which individuals find themselves, and does not exist outside of being created and reproduced by humans. No attribute is discrediting per se: only social structures make it so. Thus, an attribute such as being a virgin is stigmatized in some contexts, while in other contexts *not* being a virgin is stigmatized.

Link and Phelan (2001) construct an influential model of stigma which aims to provide a systematic framework. They argue that stigma has various essential features. First, it requires an in group and an out group: an 'us' and a 'them'. Stigma requires picking out some human differences as significant, and labelling them, but a distinguishing feature will only become stigmatizing if it is associated (or comes to be associated with) existing negative attributes. One way that this may happen is by tapping into existing bodies of judgement and assumption that have social currency (stereotypes). Often such stereotypes, despite being culturally mediated, are fairly stable across cultures.[15]

In order for some individuals to be stigmatized, they need to be considered as 'other'. The prior distinction and labelling, and association with negative stereotypes, is used to effect this. Those labelled in this way experience status loss (Link and Phelan 2001: 367). The relationship between stigma and status loss is bidirectional: stigma itself causes status loss, but low or diminished social status can itself be a cause of further stigma or discrimination (Link and Phelan 2001: 373). As stigmatization depends on the power to marshal social resources to reinforce discrimination, only groups with power are able to inaugurate stigma. For example, patients in a secure mental health setting may label the staff who treat them, and judge them according to stereotypes, and do everything else that has been mentioned so far, but it is unlikely that the staff would end up as a stigmatized group, as the 'patients simply do not possess the social, cultural, economic, and political power to imbue their cognitions about staff with serious discriminatory consequences' (Link and Phelan 2001: 376).

Stigma is thus by its nature a kind of othering of less powerful groups by more powerful groups. Power inequalities easily lead to stigmas being created where they previously did not exist; and lack of social power is one of the things that is itself often stigmatized. Stigma can lead to status loss and discrimination in three

---

[15] For example, Pescosolido et al. identify what they describe as the 'Backbone' of mental health stigma through a cross-cultural study, which found that: 'Even in countries with more accepting cultural climates, issues that deal primarily with intimate settings (the family), vulnerable groups (children), or self-harm elicit the greatest amount of negative response. A secondary core targets the unwillingness to see individuals with mental illness in positions of authority or power (work supervisors, public officials) and uneasiness about how to interact with or whether to fear violence from individuals with mental health problems' (Pescosolido et al. 2013: 858).

main ways. First, individual discrimination—as when an employer will not consider an ex-convict for a job. In this case, the stigmatizer treats another less favourably simply because of the stigmatized characteristic. Second, there is structural discrimination: in such cases, without particular individuals attempting to discriminate, people who are stigmatized are made significantly worse off by institutional structures. Third, self-stigmatization, where an individual comes to judge themselves according to the standards of those stigmatizing them (Link and Phelan 2001: 372-4).[16]

As a result of these features, stigma magnifies health-related inequalities in various ways, and is a fundamental cause of health inequalities (Hatzenbuehler et al. 2013). A sense of shame or being stigmatized will affect individuals' willingness to engage with healthcare, to disclose their conditions to others, or to continue to maintain an outgoing focus (Hutchinson and Dhairyawan 2017). Many long chronic illnesses are subject to stigma, and the fact of being stigmatized significantly worsens the experience of having a long-term condition. The fact that stigmatization arises from systematic differences in power, and further magnifies these power differentials, means that attempting to remove stigmatization without changing the underlying power structures is difficult, and can easily backfire: 'as long as dominant groups sustain their view of stigmatized persons, decreasing the use of one mechanism through which disadvantage can be accomplished simultaneously creates the impetus to increase the use of another' (Link and Phelan 2001: 375).

As Gergel (2014) argues, to the extent that policymakers' attempts to combat mental health stigma leave underlying power relationships intact, they may place the stigmatized in a double bind. If policymakers aim to promote inclusion for people with mental health problems by arguing that people with mental illnesses, such as voice hearing or schizophrenia, are *different*, then this seems to open up a set of opportunities for stigmatization in as much as the person with mental illness is framed as 'other'. In the light of the cross-cultural stereotypes revealed by Pescosolido et al. (2013), such attempts are likely to find fertile ground on which to grow. But to the extent that the person with mental illness is framed as just like 'us', then that seems to create equal and opposite problems: if the person with

---

[16] Scambler and Hopkins explore this point further by distinguishing *enacted* and *felt* stigma. Enacted stigma is *acts* of discrimination 'on the grounds of... perceived unacceptability of inferiority', while felt stigma 'refers primarily to the fear of enacted stigma, but also a feeling of shame' (Scambler and Hopkins 1986: 33) associated with having a stigmatized condition. While it was previously often assumed that it was *enacted* stigma that was the bigger problem, after extensive interviews Scambler and Hopkins concluded that, in the case of epilepsy at least, it was *felt* stigma that was the bigger problem. Less than one in three of their interviewees could recall any occasions where they had suffered enacted stigma, but nearly all experienced a keen sense of felt stigma. The implication is that felt stigma can continue to exist even in the decline or absence of enacted stigma. Scambler and Hopkins discuss the role of 'stigma coaches'—often well-meaning parents—who drill into their children at an impressionable age that certain features are shameful and to be kept hidden, even though it may be that *if* these features were revealed it would not be so bad after all (Goffman 1963: 7).

mental illness is just like us, then their behaviour could look wilful; it looks like the person should be able to act differently from how they do. Gergel thus argues that those with mental illnesses can often be subject simultaneously to two equal and opposite stigmas: likeness-based and unlikeness-based stigma. Unlikeness-based stigma is based on the mentally ill person as other 'with mental illness seen as making people intrinsically different, somehow "alien" and thus easily feared, ridiculed or restrained' (Gergel 2014: 149). Likeness-based stigma starts from the idea that the mentally ill person is fundamentally similar, but views mental illness as 'infirmity of character rather than legitimate illness' (Gergel 2014: 149).

The disquieting conclusion is that unless the underlying power structure disparities are tackled, attempts to reduce stigma are likely to suffer from what Chapter 4 described as *policy resistance*.[17] While health-related inequalities that are caused by or exacerbated by stigma can sometimes be reduced by interventions that are targeted at those individuals who are stigmatized, rectifying structural drivers of inequality will in general require structural solutions. Given the structural features that reproduce stigma we should not expect this to be easy.

The ethical case for action is strong to break up not only the stigmas that are significant causes of ill health, but also the power differentials between groups that hold them in place. Mental health stigma functions as what Wolff and De-Shalit (2007) describe as a *corrosive disadvantage*, that is, a factor that is not only unwelcome in itself but likely to cause or catalyse other unwelcome changes in the system. Mental health stigma limits and blights the kinds of lives that individuals are able to live; the commitment to treating individuals as equals requires ameliorating such suffering. However, moral clarity at an idealized level may not easily translate into workable ways of improving lives—especially where, as in the case of direct theories, no connections are forged between the principles of justice espoused and the processes by which disadvantage is maintained or exacerbated.

Where patterns of risk and disadvantage arise from intersecting structural features—as is the case with mental health stigma—they are likely to be particularly difficult to ameliorate fully. If what would be required to make a significant reduction in a particular health inequity is a set of structural interventions that go deep, then such interventions may succeed at best only partially, despite being well-designed and diligently pursued—especially if the structural features that maintain the corrosive disadvantage serve the interests of some of those who hold significant power.[18] This is, of course, not a reason not to try, but rather an

---

[17] To recall, policy resistance occurs where an attempt to intervene in a system through a 'quick fix' ends up not working because of a failure to anticipate its systemic effects.

[18] As Parker and Aggleton (2003: 18) argue, 'To untie the threads of stigmatization and discrimination that bind those who are subjected to it, is to call into question the very structures of equality and inequality in any social setting—and to the extent that all known societies are structured on the basis of multiple (though not necessarily the same) forms of hierarchy and inequality, to call this structure into question is to call into question the most basic principles of social life.'

acknowledgement of the importance of building political coalitions that are not only sustainable, but act to reinforce themselves over time, if structural justice is to be ensured.

9.7 Conclusion

This chapter has provided an account of the relationship between being treated as an equal, justice, and health-related inequalities. Given the sheer variety of different health-related inequalities, the different ways of measuring these, and the causal interactions between them, we have seen that it is crucial to clarify which inequalities are inequitable. Much of the philosophical literature has attempted to answer this question in a direct way that presupposes that justice consists in static patterns of distribution, in which everyone has the right amount of whatever it is that matters for justice. From a complexity perspective, this static vision of justice is mistaken, and so the chapter proposed an indirect conception of justice, which considers justice to be about getting institutional structures and social processes right. Framing things in complexity terms allowed us to acknowledge a fundamental difficulty for making significant progress in addressing health inequities— the ways in which many of them are held in place by structural features. The final chapter takes these ideas further in the context of communicable disease.

10

Communicable Disease

10.1 Introduction

Anyone who carries a communicable disease presents a direct threat to others. This means that even those who espouse the view that Chapter 6 critiqued as the 'autonomy first' approach will tend to agree that it is part of the role of the state to control communicable disease. For example, some libertarians argue in favour of mandatory vaccination on the grounds that if someone who could be vaccinated chooses not to be vaccinated, this imposes an unjustifiable risk of harm onto others (Flanigan 2014; Brennan 2018).

Whichever particular policy choices liberal states make in controlling communicable disease, they need to consider not only the perspectives of those persons who wish to be protected from harm, but also those whose lives will be interfered with in the interests of protecting others from harm (Battin et al. 2008). The right to public health approach provides the best way of thinking about how to reconcile these different interests. In order to see why, it is worth briefly revisiting and further specifying the discussion of John Stuart Mill's harm principle from Chapter 6.1. Mill argued that whether behaviour causes harm to others is crucial in determining if it is legitimate for the state to use the criminal law to regulate it—though, as we noted, the harm principle is now frequently invoked in a much broader range of cases than Mill initially intended.

The harm principle might look initially promising as a basis for control of communicable disease in a liberal society, given its ecumenical appeal as a normative principle. However, the appearance of the useful normative guidance is deceptive. Mill's argument in *On Liberty* aimed to rule out certain kinds of non-harm-based reasons for state action—in particular paternalistic and moralistic laws. The harm principle thus rules out certain kinds of reasons for state action, but does not by itself provide a positive account of when it is legitimate for the state to intervene to prevent harm to others. The crucial question is not whether, but when and how, it is legitimate for the state to intervene to prevent harm to others through communicable disease.

Attempts to specify the kinds of communicable disease-based harm that make state action appropriate will quickly encounter—and have to respond to—the complexity of causation of harm in communicable diseases. Communicable diseases differ in terms of their contagiousness and likelihood of severe morbidity or mortality. A cold, or a verruca, is a rather different proposition from tuberculosis or HIV. Being exposed to a pathogen is rarely sufficient for getting a clinically relevant infection with the disease, so understanding host susceptibility as well as disease-specific factors is vital.

Any successful approach to communicable disease policy that is to be inspired by Mill's harm principle would need to specify whether the goal should be to reduce all communicable disease harms, or some subset of these, such as wrongful harms. If a broad account of harms as setbacks to interest is adopted, then getting a communicable disease will usually count as a harm, and transmitting a communicable disease to others will involve being at least partially causally responsible for a harm (Feinberg 1984: ch. 1). However, as communicable disease by its nature causes harm to others, a principle telling us that the state should get involved only to prevent harm to others lacks the specificity required for useful guidance in communicable disease policy.

If, alternatively, it is specified that it is *wrongful* harms that justify state interference, then elucidation of the concept of wrongful harm in the context of disease transmission would become paramount. Given the self-reinforcing nature of communicable disease, it would require a complex combination of disease modelling, plus judgements about the substantive responsibility of others, to work out what the wrongful harm is that one particular person does in passing on a disease. Suppose person X causes person Y to become infected, and then person Y passes the infection on to several others, each of whom passes the infection on to still others. Is the harm X causes limited to infecting Y, or does X bear a share of substantive responsibility for all the harms that result from the initial infection?

Moreover, while there are cases in which individuals intentionally infect others, these cases are rare. Individuals will often not be in a good position to know whether they are infectious: many communicable diseases are infectious for a period of time before the onset of clinical symptoms; and even where the individual has experienced symptoms, the relationship between infectiousness and clinical symptoms can be complex and ambiguous. Genital herpes provides a vivid example. It is a common condition, but one that often goes unnoticed: '[a]round 80% of those affected are unaware, but the majority can be taught to recognize symptoms' (Dunphy 2014: 793). Even if an individual is aware that they have the disease, controlling their infectivity is still challenging. The rash associated with the disease comes and goes, but

> carriers shed virus in the absence of a rash for around the same number of days each year as they have a rash and...most transmissions occur when there are no lesions evident. Crucially, we have no clue as to which are the infectious days. In other words, the blistering rash, which is a badge of infectivity, can only be used to avoid half the infectious days in any set period of time.
>
> (Dunphy 2014: 794)

Any attempt to import ideas of culpability into the transmission of communicable disease needs to factor in the effects of such responsibilization on patterns of care-seeking and the likelihood of further disadvantaging some of those who are already among the worst off. Even in the case of HIV, there are few experts who think that criminalization should play more than a marginal role (Global Commission on HIV and the Law 2012). Just as in some of the noncommunicable disease cases examined in Chapter 8, thematizing the role of personal responsibility in communicable disease is likely to end up moralistic, stigmatizing, victim-blaming, or otherwise counterproductive. The nature of the harms caused by communicable diseases militate against approaches that place heavy emphasis on individual behaviour, and render more appropriate a systems-based approach to harm reduction.[1]

In short, the harm principle is likely to end up either being unhelpfully vague in the context of communicable disease, or to require specification through exactly the kind of normative work—specifying rights, their proportionate protection, and prioritization of claims—that was undertaken in Part II, and in respect of responsibility, in Chapter 8. So it does not provide an alternative to the approach that has been adopted in this book.

10.2 Thinking Ecologically About Disease

Adopting a complex systems approach reveals that the distinction between communicable and noncommunicable disease is less absolute than it might first appear. Clearly, communicable diseases can themselves be transmitted from one person to another, while noncommunicable diseases cannot. However, many risk factors for diseases are communicable, regardless of whether the disease itself is communicable or noncommunicable. Indeed, on complex systems approaches, it is entirely to be expected that what happens in one part of a system will affect stocks and flows elsewhere in the system—and thus that systemic effects will either magnify or reduce risks to health.

Models of uptake of behaviours provide a useful example. Mass advertising can be thought of as analogous to the spreading of disease spores: in one case, a

---

[1] This is not to say that individuals should not think about their own responsibility in respect of communicable disease—but just that any appeals to personal responsibility need to be handled extremely carefully. When looked at from the perspective of the individual, the question becomes one of what counts as taking adequate precautions; where the answer to that question needs to take into account both the possibility that any given person will be infectious but asymptomatic, and that different types of precautions will be more or less effective, and also that hosts will differ widely in their susceptibility to the disease. What counts as adequate precautions will differ significantly according to context: what would be reasonable precautions interacting with someone who is immunocompromised, would be very different from reasonable precautions for interacting with a healthy adult.

message is broadcast that will cause susceptible people to adopt a particular behaviour; in the other, pathogens are spread from a source that will cause susceptible hosts to become infected. Once some have been 'infected' by the new behaviour—whether it be tobacco use, or car driving, or veganism, the behaviour is liable to spread within peer groups via emulation and persuasion, with others either seeing the benefits or not wanting to be left out.[2] If the new behaviour becomes established it will alter other behaviours—perhaps becoming something that is expected and needs to be accommodated as a default, or something that people feel they need to explain if they don't do. Processes of adoption, spread, and decline of behaviours that affect health can be analysed using exactly the same models that are used for communicable disease (Page 2018: ch. 11).

While there are deep commonalities between risk factors for communicable and noncommunicable disease, communicable diseases also raise a number of distinctive questions from an ethics and policy perspective. Communicable diseases are themselves organisms, which means that they are most helpfully understood ecologically. This creates a number of challenges and opportunities that do not occur (or at least not in anything like the same way) for noncommunicable disease. Bacteria and viruses can multiply extremely quickly, and thus can rapidly cause high rates of morbidity and mortality, making interventions that control the rate of transmission a crucial component of public health policy. It is impossible to think about communicable disease sensibly without considering the dynamics of disease transmission.

Rigorous approaches must be grounded in the mathematics of transmission. The basic reproductive number, R_0 measures the average number of secondary cases that one primary case will lead to in an entirely susceptible population. R_0 depends on the length of the period during which a person is infectious, the number of contacts of the relevant type, and the likelihood a contact of the relevant type will lead to infection.[3] We can distinguish R_0, the basic reproductive

---

[2] Adam Smith's analysis in *The Theory of Moral Sentiments* of the ways in which inequalities of social power affect the spread of customs still provides a useful starting point: 'It is from our disposition to admire, and consequently to imitate, the rich and the great, that they are enabled to set, or to lead, what is called the fashion. Their dress is the fashionable dress; the language of their conversation, the fashionable style; their air and deportment, the fashionable behaviour. Even their vices and follies are fashionable; and the greater part of men are proud to imitate and resemble them in the very qualities which dishonour and degrade them' (Smith 1982 [1790]: part I, section III, ch. III).

[3] R_0 values for particular diseases are sometimes reported—particularly in the popular press—as if they were simple and objective and could be established once and for all. However, as Delamater et al. (2019) argue, this is to make two mistakes. First, 'R_0 values are nearly always estimated from mathematical models, and the estimated values are dependent on numerous decisions made in the modeling process.' Second, the value of R_0 should shift if the underlying variables that are used within models to calculate it alter from one context or time period to another. The widely quoted figure of the R_0 of measles as being in the range of 12–18 is based on data from between 1912 and 1928 in the USA, and between 1944 and 1979 in England and Wales. This is problematic given that since these times, 'major changes that have occurred in how humans organize themselves both socially and geographically make these historic values extremely unlikely to match present day epidemiologic realities. Behavioral changes undoubtedly have altered contact rates, which are a key component of R0 calculations' (Delamater et al. 2019: 3).

number, from R, the effective reproductive number. R takes into account the percentage of the population that is susceptible to infection with a particular disease. The effective reproductive rate will be lower than R_0 if any in the population either have innate immunity to the disease, or acquire immunity via vaccination or recover from it, or where social distancing measures are in place. Where R is greater than 1, the incidence of infections will rise over time; where R is less than 1, the number of infections will fall; and if R hovers around 1, then the disease will become endemic and the number of infections will stay roughly constant.

Herd immunity occurs where the percentage of those in a population who are susceptible is sufficiently low that R is less than 1. Where there is herd immunity, the rate of incidence for the disease will be sufficiently reduced that there will be a good measure of protection even for those who are susceptible. The percentage of the population that needs to have immunity—whether through vaccination or recovery from the disease—in order to obtain herd immunity is governed by R_0, and is often calculated via the simple formula $(R_0-1)/R_0$. Thus, if R_0 is 4, then three quarters of the population need to be immune for herd immunity to occur; but if R_0 is 15, then fourteen fifteenths (93 per cent) of the population will need to be immune.

R_0 and R are very simple constructs, but are still useful for thinking at a gross level about disease control scenarios. It is important to notice that R_0 and R presuppose a homogenous mixing of the population and also assume that the number of contacts per person will be constant (Fox et al. 1971). Clearly, this is a simplification that can lead to misleading results if not treated judiciously. In real-life scenarios, clustering and numbers of contacts make a significant difference: a disease will spread through a million people in a crowded city much more readily than a million people in widely dispersed villages with little contact between them. Even within the same population, how the space is used can make a great difference. For example, the effect of school closures on the transmission of influenza has been much studied (Jackson et al. 2013) and the differential effects of closure of schools and leisure activities were an important element of Covid-19 responses. Even if there is sufficient immunity within a population on average to ensure herd immunity, there may still be pockets where disease transmission can occur if cases cluster where either the rate of contacts is higher or the level of immunity is significantly lower. So, effective models that will provide useful guidance for communicable disease policy will need to consider a significantly wider array of variables, including vaccine effectiveness and subgroup differences (Fine et al. 2011).

More sophisticated models of communicable disease will take into account a greater variety of features of hosts, pathogens, and environments, and the relationships between these. As in other cases of systemic causation, reducing or controlling prevalence may require us to look to features that are initially far from obvious. Ostfeld (2012) provides a useful and detailed complex systems analysis in the case of Lyme disease, which also gives some pointers for the kinds of inter-actions that may be present in other communicable diseases. As Ostfeld and others have painstakingly documented, in North America, Lyme disease infec-tion occurs when a susceptible human is bitten by a blacklegged tick. The tick itself is merely the disease vector; it is the *Borrelia burgdorferi* spirochete that is the cause of the disease in humans.

Blacklegged ticks are not born infected with *B. burgdorferi*; each one must acquire it by feeding on an infected host. The life cycle of blacklegged ticks goes through four stages. Eggs hatch into larvae. Each larva will aim to attach itself to a suitable host such as a mouse or shrew for a blood meal. Once the larva has had its fill, it drops off the host and remains on the forest floor until the following spring, when it transforms into a nymph and, later, an adult. Ticks can become infected by feeding on an infected host as larvae, but are more likely to become infected as nymphs.

The likelihood of each tick becoming infected depends both on whether the host it feeds on is itself infected with *B. burgdorferi*, and also on the competence of the host at transmitting the disease. Ostfeld and others' studies have shown a very wide degree of difference in competence of reservoirs, with the implication that while 'four species of rodents and shrews is feeding nearly half of all the lar-vae in the forest, and infecting about 90% of the ones that get infected, all the other hosts are feeding more than half of the larvae and infecting only 10% of them' (Ostfeld 2012: 58). Given the systemic nature of communicable disease, this has important implications for human infection rates over time: fewer larval ticks becoming infected with *B. burgdorferi* will mean fewer transmissions of *B. burg-dorferi* to another host, and this process will iterate over generations.

Properly mapping this system allows us to see some potential solutions that would otherwise have seemed wildly counterintuitive. For example, given the vital importance of competent host reservoirs in Lyme disease transmission, an unexpectedly significant question is what kinds of environments are conducive to the flourishing of species such as opossums, which have very low competence as a host of *B. burgdorferi*, but which will be fed on by many ticks. (A core part of the problem, according to Ostfeld, is that in the areas of the USA that have been worst affected by Lyme disease, woodlands have been parcelled up into relatively small chunks alongside housing, and these chunks are large enough to support small mammals such as shrews and mice that are competent hosts, but not the larger mammals that are less-competent hosts. If so, increasing the size of woodlands might make a significant difference to rates of Lyme disease infection.)

Looking in detail at the systems within which communicable diseases grow and thrive and infect human beings gives strong evidence that ecological approaches, which take account not just of the nature of the relevant pathogens, vectors, and environment, but also take seriously the interactions between them, will be more fruitful than simpler or more reductive approaches.

Of course, such an approach must be data-driven. Just as in other cases of mapping complex systems, it is not enough to make loose claims about everything being connected to everything else; rather, the nature and the strengths of these connections needs to be explored empirically (Ostfeld 2012: 186). Such knowledge is likely to be contextual: detailed understanding of the ecology of one communicable disease may not imply actionable insights for other communicable diseases that occupy different ecological niches.[4] Doing so will require a combination of detailed empirical work, with a nuanced ethical understanding of the advantages and disadvantages of intervening at different points within the system—something which will require a level of empirical detail and specificity beyond what we can provide here. However, the broad contours of taking an ecological approach to communicable disease is something we shall explore for the rest of this chapter.

We will focus on three specific problems. First, fair usage of antibiotics. Drugs can become ineffective as pathogens evolve resistance to them, raising difficult questions of what counts as a fair strategy to prioritize their usage (Section 10.3). Second, maintaining herd immunity: successful vaccination policy for many communicable diseases requires the maintenance of high rates of immunization in the community, which is itself dependent on maintaining public confidence in vaccines (Section 10.4). Third, eradication of disease: it is possible to globally eradicate some communicable diseases by destroying either the organisms that cause the disease or their vectors—raising important questions about how to weigh up the tremendous long-term health benefits that can come from successful eradication programmes against other seemingly more urgent priorities (Section 10.5).

10.3 Fair Use of Antibiotics

Antibiotics are drugs that promote recovery from infections by either killing bacteria or inhibiting their growth. There are several different classes of antibiotics, which differ in chemical structure, mode of action, and the bacteria against which they are effective. It is common to distinguish between narrow and wide spectrum antibiotics: narrow spectrum antibiotics are effective against a small range of bacteria, while wide spectrum antibiotics are effective against a broader range of bacteria.

Bacteria develop drug resistance when specific antibiotics lose their ability to kill or inhibit the growth of these bacteria. Given the variety of antibiotics, bacteria can be resistant to some but not all antibiotics. Some antibiotic resistance is inherent: that is, some bacteria were already resistant to some antibiotics at the

---

[4] However, there is a growing body of evidence that loss of biodiversity may be a more general cause of rises in communicable disease infections in humans. On this point, see Keesing et al. (2010).

time when these antibiotics were discovered. Bacteria have existed as part of a Darwinian struggle for life for over three billion years; and those that have survived this far have already faced selection pressures in which the ability to resist naturally occurring antibiotics has increased reproductive fitness (D'Costa et al. 2011). For example, some bacteria have developed efflux pumps—and so automatically pump antibiotics out of their cells before the antibiotics have a chance to kill them.

Other antibiotic resistance is acquired: this occurs when bacteria that were previously susceptible to antibiotics become resistant to them as a result of genetic changes: for example, by modifications of the target site or metabolic pathway that the antibiotic had previously made use of. Drug resistance spreads by two means: vertical gene transfer, and horizontal gene transfer. Vertical gene transfer is the familiar process of natural selection. A random point mutation occurs in the process of cell division, which is passed on to the next generation. Horizontal gene transfers—whose importance for evolution more generally has only been fully appreciated since the 1980s—occur when an organism incorporates genetic material from cells other than the bacterium's ancestors.[5] Horizontal gene transfers allow drug resistance genes to be passed between species of bacteria through the exchange of plasmids. The vast majority of mutations and horizontal gene transfers will not be advantageous to the organism; but where there are billions of bacteria, and a very large number of generations, mutations that provide an advantage in the struggle for survival will be selected for over time.[6]

Current usage of antibiotics is unsustainable. That is to say, the rate at which new classes of effective antibiotics are being brought to market is much lower than the rate at which existing classes of antibiotics are becoming ineffective due to resistant pathogens. We are facing the threat of a 'post-antibiotic age'. To give some scale of the current problem, drug resistant infections currently account for around 700,000 deaths per year, and this could grow to 10 million per year by 2050 if nothing is done (Review on Antimicrobial Resistance 2014).

A number of policy responses have been proposed or partly implemented, including altering the balance of funding mechanisms, so that more resources are devoted to ensuring that new classes of effective antibiotics are brought to market; preventing antibiotics being used routinely for weight gain in farming; and quicker testing so that antibiotics can be more precisely targeted and the need to use wide spectrum antibiotics on an empirical basis is obviated. However, there remain challenges, at technical, institutional, and political levels in changing

---

[5] For an overview of horizontal gene transfer, see Syvanen (2012), or Quammen (2018) for a longer and more accessible account.

[6] Oddly, there has sometimes been a reluctance to explain drug resistance in clear evolutionary terms (Antonovics et al. 2007), but the basic mechanism is that of natural selection.

patterns of antibiotic usage and discovery quickly enough to avoid grave increases in mortality.[7]

Without new classes of antibiotics, severe and worsening problems of resistance are to be expected. Introducing a new class of antibiotics will give us some breathing room, but given the nature of microbes and of natural selection, there is a high likelihood that any new class of antibiotics will itself become ineffective over time as resistance to it spreads. Given the continuing rise in global population (to at least 9.5 billion by 2050), and the fact that many are still suffering avoidable morbidity and mortality through lack of access to antibiotics, demand for therapeutic antibiotics may continue to rise. It is plausible (though not inevitable) that, despite the best efforts of our future research, using antibiotics in healthcare practice in the way that is now thought to be best practice will be unsustainable (Millar 2011), and so developing an approach to the fair usage of antibiotics is crucial.

10.3.1 Static Models for Fair Distribution of Antimicrobial Effectiveness

It is important to distinguish two types of resources in thinking about antibiotics. First, physical stocks of antibiotics in the form of tablets or vials, and second, stocks of the more abstract good of *antibiotic effectiveness* (Wilson 2013). When antibiotics were first discovered, there was significant difficulty in manufacturing enough of them to treat all those with susceptible infections. Most antibiotics are now off patent, and can be manufactured very cheaply,[8] but it is much more expensive and difficult to create more antibiotic effectiveness. Indeed, the cheapness of antibiotics has contributed significantly to the problem of scarcity of antibiotic effectiveness, for example by making it economically advantageous to use antibiotics in animal feed.

Antibiotic resistance is thus a problem of scarcity of the *effectiveness* of antibiotics, rather than scarcity of stocks of antibiotics themselves. Once we clarify that the resource that we are trying to distribute fairly is antibiotic effectiveness, and that this is scarce, then the key question is what it would be to respond equitably to this scarcity. I shall explore two different kinds of models for fair distribution: static models and dynamic systems theoretic models, arguing that only dynamic models are adequate to the task.

---

[7] For an overview of the ethical issues raised by antimicrobial resistance, see Littmann and Viens (2015).

[8] This is not to say that antibiotics are always available in resource-poor settings: under-supply as well as over-supply is a problem.

One approach to the distribution of antimicrobial effectiveness would be to treat it as a variant on a traditional resource allocation problem, along the lines of the competing claims approach that was examined in Chapter 7. For example, antibiotic effectiveness could be thought of as a scarce resource to which different individuals have a claim—where the strength of this claim depends on features such as the severity of their illness, and how much they stand to benefit.

One important feature that distinguishes drug resistant infections from the thought experiments in which the competing claims account is usually deployed is the depth of the uncertainty. If policymakers respond to the antibiotic crisis by rationing access to antibiotics, this will require individuals to forgo treatments now in order to prevent things getting worse in the future. Each individual who is denied access to an effective antibiotic now could reason as follows: (1) I am ill now and require treatment; (2) it is far from clear whether any particular individual will benefit from my forgoing the good now; (3) quite irrespective of what I do, the drug in question may become either unusable in the medium term (because of the rise of drug resistance), or unnecessary (because of the invention of an alternative, such as a bacteriophage, or a new class of antibiotics); (4) therefore it is unreasonable to think that my claim to receive the antibiotic is weaker than the claims against which I am (or could be) competing.

This makes pairwise comparison significantly more difficult to apply over time than in resource allocation cases where effectiveness will remain constant. If we attempt to make a comparison between the strength of claim that someone has now to antibiotic effectiveness, and the claim that someone, perhaps years in the future will have to the effectiveness of that antibiotic, then it is difficult to see how there could be an objectively correct answer; and in weighing this up now, we face an obvious conflict of interest. To focus on just one question, if whatever policymakers decide to do now, this class of antibiotic will be less effective in the future than it is now (because others continue to use it), how does that affect the claims of those in the future to still be able to gain access to it?

One answer might be just to apply a discount rate to the strength of the claim— for example, to model the benefits that the antibiotic will be able to provide in ten years' time, and use that as a discount rate.[9] However, in so doing we run up against the challenge of performativity that was discussed in Chapter 4, as the key factor controlling how effective antibiotics will be in the future is how they are used in the intervening time. It would be obviously unfair if we were to predict that a particular antibiotic will become completely ineffective because we will use it so much, and then use that as a reason for treating the claims of individuals in the future to be able to use it as negligible.[10]

---

[9] I discuss time discounting in more detail in Chapter 7.6.

[10] There are interesting comparisons to be drawn between drug resistance and climate change. In both cases, getting a grip on the problem requires those making decisions now to take the interests of

10.3.2 Thinking Dynamically About Fairness and Antibiotics

There are further reasons for suspicion about a static distributive model of antimicrobial effectiveness, which become apparent once we adopt an ecological perspective. Competing claims accounts of the allocation of scarce resources are usually able to make the assumption that the amount of the resource to be distributed is constant—and that what needs to be determined is how to distribute fairly the fixed supply of the resource. For example, funding a new intervention programme will mean that there will be less money for existing programmes. However, an accurate understanding of antibiotic effectiveness requires us to view its quantity as variable and to some extent unpredictable, rather than static. Not every case of usage of antibiotics will lead to the production of resistant strains; and not every production of resistant strains will be transmitted onward. Developing new classes of antibiotics increases the stock of antibiotic effectiveness available to us.

By itself, this would imply only that antibiotic effectiveness is affected to some extent by randomness, and that it may rise and fall. Other additional features require us to think of the stock of usable antibiotic effectiveness as performative and relational. If resistance traits do not confer a selection advantage in a particular context, then resistant bacteria may be out-competed. Conversely, there can be mutations separate from anything that human beings do, which reduce the fitness costs of resistant traits, and make them more likely to persist even where no antibiotics are being used. Given that different strains of bacteria will be resistant to different drugs, the amount of antibiotic effectiveness will alter also simply by moving drugs around from one location to another. Thus, the amount of antibiotic effectiveness is a dependent rather than an independent variable: the amount of it will depend on the total state of the system—rather than existing separately from it.

As also became apparent from looking at Lyme disease, understanding drug resistance ecologically can provide surprising and counterintuitive ways of reducing the speed of the spread of drug resistance. The work of Andrew Read and his collaborators is emblematic in this regard. Kennedy and Read (2017) explore an interesting question: why does drug resistance develop much more readily than

---

those in the future sufficiently seriously to avoid vicious circles of short-termism. This is much easier said than done. Levin et al. describe climate change as a super-wicked problem, noting 'the tendency of our political institutions to make decisions that give greater weight to society's immediate policy interests and to delay required behavioral changes, often by the use of ostensible commitments to reduce greenhouse gas emissions that have little or no immediate effect. These tendencies to punt policy choices are exacerbated by near-term costs and the beliefs by some that these disincentives will diminish in the future (e.g. via the availability of less expensive technology or increased costs of inaction). Yet as the future approaches—when we planned to act on climate change—the salience of the short-term costs returns, presenting a vicious cycle' Levin et al. (2012: 128–9). Littmann et al. (2019) argue that this super-wicked problem framing should be applied to antimicrobial resistance.

vaccine resistance? Both vaccines and antibiotics are deployed against living pathogens, and would be expected to create selection pressure in favour of resistance. But yet while drug resistance is common and deeply problematic, vaccine resistance is much rarer, and when it does occur the consequences for human health are much less severe (Kennedy and Read 2018). They theorize that there are two key features that make the difference. First, vaccines are prophylactic, aiming to prime the immune system before the pathogen is encountered, whereas antibiotics are prescribed only after there is a clinically relevant infection. By the time that an infection has got bad enough that someone is prescribed antibiotics, there will be orders of magnitude more pathogen particles present than when the pathogen is first encountered by the primed immune system. So, the selection pressure created by the vaccine is much smaller than that created by the antibiotic.

Second, there are great differences in the fragility of the mechanism used. Vaccines typically prime the immune system in such a way that it is responsive to many features of the pathogen. If one or two of these features change, this will not make any difference to its effectiveness. However, the mode of action of antibiotics is usually much less resilient to changes in bacteria: it may be that only one or two crucial mutations will greatly lower the effectiveness of an antibiotic.[11]

Thinking through the implications of selection pressure should lead us to think about fairness in the use and supply of antibiotics in a more sophisticated way. At a deep level, the problem is not one of allocating a fixed scarce resource fairly, but how to design institutions that reduce likelihoods of infections, and how to design interventions that will minimize the chances that they will become ineffective through use.[12] Questions about the fair use of antibiotics, considered as a scarce resource, should be seen as secondary to questions about how to reduce the incidence of infections that require antibiotics as a remedy. In this vein, it has been argued that a new generation of evolutionary ecology-inspired strategies is required, which will act 'not necessarily to cure the individual patient...but to "cure" specific environments from AbR [antibiotic resistance] and to prevent or weaken the biological elements involved in AbR' (Baquero et al. 2011: 3650).

Where the need for antibiotics can be obviated by interventions such as vaccination, or by attacking the conditions that lead to weakened immune systems in the first place, we should do so. Vaccines can help to reduce usage of antibiotics

---

[11] Notwithstanding the slight effects of resistance, the protection provided by vaccines may be relatively short-lived. The immune response created by a vaccine will often wane over time, requiring boosting. In addition, mutation, or competition between different strains of viruses, may render less effective the immune response created by a vaccine (hence the need to reformulate influenza vaccines each year as different strains come to dominate). Such shifts within strains of the virus should be understood in terms of selection pressure; but within this selection pressure, it is factors other than the use of vaccines which predominate.

[12] In Millar's memorable image, 'Antibiotics are like a leaky rescue craft which becomes less effective the more it is used' (Millar 2012: 465). We need a rescue boat that is less likely to sink, and also to reduce the number of rescues required.

by reducing the circulation of susceptible diseases and establishing herd immunity, and can also be helpful in preventing unnecessary prescriptions (Lipsitch and Siber 2016; Jansen et al. 2018).[13] Mapping the social determinants of communicable disease—and of drug resistant infections in particular—will also be crucial. Wild et al. (2017) provide a useful example of the kind of thinking that is required—showing how the structural injustice and human rights abuses suffered by vulnerable migrant populations, and lack of access to stable and accessible care, function as an important determinant of drug resistant tuberculosis.

10.4 Vaccine Hesitancy, Herd Immunity, and the Dynamics of Trust

Vaccination has prevented tens of millions of deaths, and is generally very cost-effective. Childhood vaccination in particular is a key public health strategy. It provides an exceptionally cost-effective way of reducing or removing health risks from the environment; it does so safely, and with little or no inconvenience to those vaccinated. Where herd immunity can be ensured, it also provides effective ways of protecting the most vulnerable, and those who cannot be vaccinated themselves. There are strong reasons for thinking that failure to provide an adequate and easily accessible schedule of childhood immunization would violate the right to public health, as Chapter 6 argued.

Thinking in systems terms alerts us to the fact that interventions that decrease one person's risk factors may have spillover benefits to others—as when a targeted exercise programme ends up also causing the friends and colleagues of those targeted to exercise more. Vaccination is unusual in that it is often undertaken *in order* to achieve these secondary benefits.

Despite the obvious personal and public health benefits of vaccination, recent years have seen the rise of vaccine hesitancy in many parts of the world. Vaccine hesitancy is defined as 'the delay in acceptance or refusal of vaccines despite the availability of vaccination services' (SAGE [Strategic Advisory Group of Experts on Immunization] 2014), and was ranked in 2019 by the WHO as one of the ten greatest threats to global health (World Health Organization 2019a). Much concern has focused on measles vaccination. Because of its high R_0, herd immunity for measles requires an immunization rate of 90–95 per cent.[14] Achieving this

---

[13] Up to 50 per cent of US antibiotic prescriptions for acute respiratory conditions are prescribed inappropriately, according to one review (Fleming-Dutra et al. 2016). Better population coverage with influenza vaccine would significantly reduce such unnecessary prescribing (in so far as it reduces rates of respiratory infections that cause patients to be prescribed inappropriate antibiotics).

[14] The R_0 of measles is usually cited as in the range of 12–18. Applying the simple $(R_0-1)/R_0$ formula to an R_0 of 12 would imply that 91.6 per cent coverage would be required, while an R_0 of 18 would require 94.4 per cent coverage. However, as the previous section indicated, things will be more complex than this—both because the data on which these R_0 calculations were made are now quite old

rate of coverage is challenging. Additionally, vaccine hesitancy tends to cluster, leading to local outbreaks even if overall national level coverage is good. Measles vaccination matters, given both the infectiousness and the fact that measles can be deadly. WHO estimates that there were 110,000 deaths globally from measles in 2017, heavily clustered in a small number of countries that lack adequate routine immunization, and that in 2000–17, 'measles vaccination prevented an estimated 21.1 million deaths' (World Health Organization 2019b). A number of wealthy countries have seen spikes in cases in recent years.

The reasons for inadequate vaccine coverage rates include the 'three Cs' of complacency, confidence, and convenience. Complacency exists where 'perceived risks of vaccine-preventable diseases are low and immunization is not deemed a necessary preventative action' (SAGE 2014). Confidence relates to levels of public confidence in the vaccine providers as well as other actors and the politics surrounding vaccination programmes. Convenience encompasses the physical availability, geographic accessibility, and affordability of vaccines, as well as people's health literacy and ability to understand the value of immunization services (SAGE 2014). This section focuses on cases related to confidence and complacency rather than convenience—as the right to public health account is unequivocal about the requirement that childhood vaccination be convenient.[15]

Childhood vaccination programmes target those who are asymptomatic, and include in their coverage those who would be at low risk. Confidence and complacency can present important challenges. A nodding acquaintance with the literature on herd immunity may give parents the impression that, so long as enough other parents get their children vaccinated to ensure that there is herd immunity, the risk to their own child from not being vaccinated will be very low. This thought may be supplemented by the further thought that whether their child is vaccinated will make only a negligible difference to whether herd immunity can be established—and so there is nothing wrong in refusing vaccination.

Clearly, this line of thinking has performative implications. If even one in ten parents—let alone all parents—adopted this free-riding approach to vaccination, then maintaining herd immunity for measles would be extremely challenging, if not outright impossible. It is correct to think that *if* the level of circulation of the pathogen is extremely low, *and* there is some risk or inconvenience attached to the vaccination, *then* it may not be in the individual best interests of the child to be vaccinated. However, this is far from the case in scenarios such as routine childhood vaccination, where the vaccines are extremely safe, and the levels circulating in the community are such that it is still very possible to get the

and probably do not represent accurately current levels of contact rates (see footnote 3 in this chapter), and R_0 even if calculated on the basis of accurate data, presupposes homogenous mixing.

[15] This, and the next three paragraphs, draw on work with Tracey Chantler and Emilie Karafillakis. See Chantler et al. (2019).

infection. So free-riding on herd immunity by not vaccinating is unlikely to be in the child's best interests in the usual circumstances in which childhood vaccination is offered.[16]

Of course, parents may disagree about whether vaccination is in the child's best interests. There are some parents who object to vaccination on either philosophical or religious grounds, or simply take a view of the risks and benefits of routine immunization that is different from that which would be gained from a careful analysis of the relevant peer-reviewed studies.

In interrogating such claims, it is important to be clear about what parents should be trying to do in making decisions on behalf of their children. It is now generally assumed that parents should be aiming to act as a trustee of their children's interests, rather than making use of a power of parental dominion to determine their children's future as they see fit. Given this, parents must be accountable at least to some extent to others in their judgements about what constitute the child's interests. Parents who refuse to allow their children to be vaccinated may not be merely making a mistake about the best way to pursue the child's interests, but may also be breaching the ethical duties that they have to their children. There is a complex set of questions about where these limits lie and how to determine them (Birchley 2016). These questions are heightened by the rise of vaccine hesitancy, given that the purpose of childhood vaccination policies is not just to protect those children who are vaccinated, but also to reduce the spread of disease and to protect those who cannot be vaccinated.[17]

It cannot be assumed that vaccination will always be in the *ex ante* self-interest of all those vaccinated, however. As Section 10.5 examines in more detail, as a global eradication campaign moves closer to success, less and less of the expected benefits of a vaccination will accrue to the person vaccinated, and more and more to the world at large through the elimination of the health threat from the environment. As the number of cases of the disease approaches zero, the expected benefit to individuals who are vaccinated may become less than the expected costs, if the vaccine itself poses at least a minimal risk (Barrett 2013). It is sobering to realize that there were between 200 and 300 deaths in childhood as a result of complications such as encephalitis following smallpox vaccination in the USA between 1948 and 1965, but only one US death from smallpox in this period (Stepan 2011: 210). The risks of the oral polio vaccine are much smaller than those from the smallpox vaccine, but they are far from infinitesimal. It is thus not

---

[16] As Chapter 7 discussed, when making decisions under uncertainty concepts such best interests should be judged *ex ante* rather than *ex post*. The relevant question is whether it would be in the child's best interests to be vaccinated, knowing what a parent would reasonably be in a position to know at the time that the decision to vaccinate needs to be taken. Even in the very small number of cases of vaccine-related harm, vaccination will (unless it was contraindicated) have been in the child's *ex ante* best interests.

[17] In the case of tetanus, the parent's duty to pursue the child's best interest would be only for the obligation to vaccinate, given that tetanus is not communicable from person to person.

immediately clear that a global vaccine-based eradication campaign could be successfully completed if all healthcare professionals took literally the demand that each intervention they provide should be in the best interest of each patient considered as an individual.[18]

Thus, it matters whether vaccination and other disease control measures are framed within the context and ethical assumptions of standard medical treatment, or within a broader context of public health policy. In clinical care, the expectation is that the recipient of the treatment will be its main beneficiary; to give just one example, the International Code of Medical Ethics states that 'a physician shall act in the patient's best interest when providing medical care' (World Medical Association 2006). In standard vaccination campaigns, the expectation that the individual person vaccinated is the main beneficiary remains, but such campaigns also aim to create spillover benefits to others from herd immunity. Where a policy is known to be *ex ante* worse for some, such justifications cannot be run.

As Chapter 5 argued, it is unreasonable to expect that an ethically acceptable government policy that aims to promote the common good will *never* be expected to make some individuals worse off than they otherwise would be. Thus, even if it would be *ex ante* worse for some individuals to be vaccinated or to adopt other disease control measures, this would not by itself entail that such a policy is unethical. So, to the extent that eradication campaigns (or lockdown policies) are compared to ordinary medical practice they may look ethically problematic, but to the extent that they are compared to public policy contexts such as transport they may seem relatively unproblematic.

In thinking through whether it is reasonable for states to request that individuals take precautions or otherwise modify their behaviour to improve the safety and well-being of others, what matters is not the magnitude of the expected harm that would be averted by each person taking the precautions, but the proportionality between the magnitude of the risk to be averted, and the costs or burdensomeness of measures that would reduce that risk. As was discussed in Chapter 7.3, it is very implausible to argue that small additional risks make *no* ethical difference. If this were the case, then even a large number of small additional risks would make no difference.

Large, and even catastrophic, harms such as those caused by climate change and antimicrobial resistance are created by myriad events, each of which makes negligible difference. While it is unlikely that one person's vaccine hesitancy will cause someone else to get a measles inflection that will lead to their death, this is far from inconceivable. Some will not be able to be vaccinated for medical

---

[18] The adoption of disease control policies that were *ex ante* worse for some was also a salient feature of lockdown policies adopted for Covid-19: children and teenagers, who would be very unlikely to get the disease severely, had their lives and schooling disrupted by lockdown policies (John 2020).

reasons—for example, being too young, or immunocompromised. So it would be problematic to assume that vaccination should only be a matter of individual responsibility; this would be to fail to protect the most vulnerable. Given the importance of features such as protecting those who are vulnerable or otherwise worst off, it is justifiable for states to slightly increase risks of harm to some in order to protect others. What is crucial is that the risks are distributed fairly, and that the *ex ante* risks imposed on individuals are not excessive in relationship to the benefits to be obtained.

This analysis allows us to draw two conclusions. First, where vaccination would be in the *ex ante* interests of the child, and also forms part of a public health strategy designed to protect the common good, it is prima facie unreasonable for parents to refuse for their children to be vaccinated according to the routine immunization schedule. The case where it is *ex ante* worse for the child is a little more complex. Assuming one is deciding for oneself, if the burden of undergoing routine vaccination is very small in comparison both to what those unable to be vaccinated stand to lose, and where sufficient compliance with the policy will allow a significant common good to be realized, then it would also be prima facie unreasonable to refuse. Are parents, in acting as trustees for their childrens' interests, also to be expected to agree to a result that is *ex ante* slightly worse for their children, if this policy is necessary to protect others who are even worse off? Yes: the way in which parents hold their children's interests in trust presupposes that the actions taken to secure the interests of the child are themselves ethically justifiable. I shall come back to what follows from this for mandating vaccination in Section 10.6 after a deeper consideration of eradication—which poses the challenge of vaccination policies that are not in the *ex ante* interests of all in its strongest form.

10.5 Disease Eradication

Global eradication of disease has fired the imagination since the introduction of vaccination, a possibility that Thomas Jefferson brilliantly expressed in his letter to Edward Jenner in 1806: 'Medicine has never before produced any single improvement of such utility... Future nations will know by history only that the loathsome small-pox has existed and by you has been extirpated' (Jefferson 1806). While it was over 170 years before Jefferson's dream was realized, smallpox was indeed globally eradicated by the end of the 1970s, and remains an iconic achievement of the twentieth century.

In general, to eradicate a disease is to reduce to zero the incidence of the disease through deliberate efforts (Fenner et al. 1998). To eradicate a disease globally is to remove the disease threat from the whole world, permanently: in a consensus definition, 'the worldwide absence of a specific disease *agent* in nature as a result of deliberate control efforts that may be discontinued where the agent is

judged no longer to present a significant risk from extrinsic sources (e.g. smallpox)'
(Cochi and Dowdle 2011: 5).

It would be difficult to deny that the global eradication of smallpox, which had
been a major cause of morbidity and mortality for thousands of years, was a good
thing. To this extent, the ethics of eradication is straightforward. However, it is
important to counterbalance this ethical commonplace with the recognition that
there were a number of failed and expensive eradication campaigns in the twentieth century, including yellow fever, yaws, and malaria (Stepan 2011). In some
cases—like yellow fever—the disease should probably not have been a candidate
for eradication attempts in the first place, as it has an animal reservoir. In other
cases, the failure may more accurately reflect the intrinsic difficulty of globally
eradicating a disease, even where it is correctly judged to be technically feasible
to do so.

Factors responsible for this high level of difficulty include the degree of international coordination and cooperation over a prolonged period that are required
for successful global eradication campaigns, the challenges of ensuring that
enough individuals continue to be vaccinated to maintain herd immunity everywhere in the often long period between the disease being eradicated locally and
being eradicated globally, and the continual risk that cases will be exported back
into territories that were previously free of the disease as a result of war or political instability (Klepac et al. 2013). The long endgame of the polio eradication
campaign provides a vivid example. The World Health Assembly committed to
the eradication of polio in 1988, with eradication originally scheduled to be completed by the year 2000. Instability then saw an increase in the number of countries exporting wild poliovirus, and a WHO declaration of a Public Health
Emergency of International Concern in 2014.[19]

Which is the right frame to bring to the ethical consideration of eradication
policies? Eradication policies are likely to have a long 'last mile', in which millions
or even billions need to continue to be vaccinated for a prolonged period of time
to complete the task, despite the fact that there are very few remaining cases of
actual infections. The time and resources required for compliance with the
demands of an eradication programme could instead be devoted to goals that
may be more pressing in a particular state. So, where disease eradication policies
are ethically justified, this may be more plausibly done on the basis of the sheer
scale of the health benefits to be obtained, than by the claim that any individual's
right to public health would have been violated if a disease containment policy
had been pursued instead and the resources used for other pressing public
health purposes.

This section examines how to think about global eradication as a policy
goal—and in particular whether it is appropriate to think of eradication as

---

[19] This remained in force at the time of writing (December 2020).

ethically exceptional (Wilson and Hunter 2010). If there is, we should expect eradication policies to be subject to *sui generis* ethical considerations; if there is not, we should expect standard approaches to the ethics of public health policy to be sufficient. I begin by examining three arguments that have been put forward for thinking that eradication is in some way special as a policy goal. These are (1) that global eradication has symbolic importance; (2) disease eradication is a global public good, and (3) disease eradication is a form of rescue. None of these arguments succeeds in showing that eradication is *sui generis* as a policy goal. None provides a reason for thinking that public health authorities have special duties to pursue eradication campaigns, or that individuals have special duties to facilitate them.

However, the fact that these arguments fail does not entail that global disease eradication is ethically problematic, or that it should not be undertaken. Global eradication of a disease, if successful, is a way of providing an enormous health benefit that stretches far into the future. There is no need to reach for the idea that there is a special duty to eradicate disease; the same considerations that are in play in ordinary public health policy—of reducing the burden of disease equitably and efficiently—suffice to make global disease eradication a compelling goal where doing so is feasible.

10.5.1 The Symbolic Value Argument

Eradication is often thought to have an important symbolic value. The tangible goal of eradicating polio has energized donors—such as members of Rotary Clubs—for many years. As Margaret Chan, the then Director General of the WHO, put it in a speech to the Rotary International Convention in 2008, 'We have to prove the power of public health. The international community has so very few opportunities to improve this world in genuine and lasting ways. Polio eradication is one' (Chan 2008).

It is sometimes argued that this symbolic value makes eradication an ethically special case—and hence that eradication policies should be pursued over and above the actual health benefits they provide. Certainly, as we explore in more detail later, eradication policies need to stay the course, and large-scale success stories like smallpox help to make the goal seem achievable. But this is merely to say that eradication requires a firm long-term commitment *if* it is to be successful, rather than to take the symbolic value of eradication to be a reason to undertake such a policy in the first place.

Achieving a goal of high symbolic value can be energizing, and (as Margaret Chan suggested) form an important part of a narrative about the ways in which public health matters and can achieve lasting change. However, symbolic value does not negate the requirement for public health policies to be justifiable to individuals. So, even if it is agreed that eradication has a high symbolic value for many individuals, this does not entail that anyone has an additional ethical duty to facilitate eradication campaigns by agreeing to be vaccinated, or that governments have an additional permission to do things that would otherwise constitute a violation of someone's rights, such as enforcing vaccination.

If the person to be vaccinated agrees that disease eradication has high symbolic value, then it seems plausible to suppose that she would be willing to take the steps necessary in her own conduct to facilitate disease eradication, and to allow others to interfere with her life for this purpose. But the operative moral principle here is consent, and the symbolic value of eradication plays only a derivative role. The disease for which there is a global eradication campaign will often be a relatively minor cause of morbidity and mortality within the state's territory, when compared to other public health problems citizens face. If someone does not think that disease eradication has an important symbolic value, it is difficult to see how the fact that it had symbolic value *for others* could either generate a moral duty for her to subject herself to risk, or a permission for others to coerce her in order to preserve this symbolic value.

10.5.2 The Global Public Goods Argument

It is sometimes argued that global eradication of disease is a key example of a global public good—a good that is both non-excludable and non-rival: 'Once provided, no country can be prevented from enjoying a global public good, nor can any country's enjoyment of the good impinge on the consumption opportunities of other countries. When provision succeeds, global public goods make people everywhere better off' (Barrett 2007: 1).

In other contexts where public goods need to be provided, it is usually taken for granted that communities may legitimately require their members to contribute to the provision of these goods regardless of whether so doing is in the best interests of each person considered as an individual. Obvious examples would include jury service or paying one's taxes. So it might be thought that the mere fact that eradication is a global public good is sufficient to show that there are special ethical duties to undertake disease eradication policies.

However, it is unclear that disease eradication does have a status as a public good in a way that sets it apart from policies that aim at disease control rather than eradication. Risk reductions in general would plausibly appear to be public goods, as they are usually nonrival and nonexcludable. If so, the global public goods argument does nothing to support policies of risk elimination (eradication) over risk reduction (control). If global public goods theorists wish to maintain that eradication alone, and not mere risk reduction, is a global public good, then they need to explain why.

In the quoted passage, Barrett suggests that it is the *universality* of the benefit that is key, and it is this that allows Barrett to say that 'people everywhere are better off' as a result of the global public good. However, it is unclear in what sense people everywhere would benefit from the eradication of a disease such as guinea worm. The most obvious answer is that there is a current health threat that would be removed from the environment, and so everyone would be benefited by living in a safer environment. However, the degree to which the environment is made safer, and the ways in which it is made safer, and for whom, need to be specified. In this case it is unclear in what way citizens of a country that did not have guinea worm (for instance the UK) would be benefited by global eradication of the disease. Or, if this is a benefit, then it is unclear that it is a large and significant benefit for those individuals.

In addition, it would be puzzling to claim that a *risk reduction* for a particular disease is not a global public good, but an elimination of that risk is. All human beings will die at some point or other. So even if one particular disease is eradicated, it will still be the case that everyone will die of some disease or other. So while it might be possible to conceptualize the elimination of *a* threat to health as a global public good, it is unclear why we should think of the reduction of a particular risk to health to zero to be specially significant, where there are still many risks to health in the environment. In either case, the appeal to eradication as a global public good does little to justify either the claim that individuals have special duties to facilitate eradication campaigns, or that public health authorities have special permissions to pursue them.

10.5.3 Is Eradication a Form of Rescue?

Claudia Emerson argues that the duty to rescue provides the main reason to adopt plans to eradicate disease:

> The duty to rescue obliges one to rescue someone in distress provided one has the ability to do so, and doing so does not require excessive sacrifice...Consider the case of polio, where it is projected that the failure to complete eradication will result in 4 million children contracting paralytic polio over the next twenty years...Failure to eradicate in this case is synonymous with a failure to rescue, given that we have the means to save those 4 million children from the harm of polio. (Emerson 2011: 107)

It is important to distinguish between obligations of rescue and more general obligations of beneficence. Common-sense morality takes obligations of rescue to be much more stringent than those of beneficence. Rescue cases involve identifiable individuals who are in peril *now*. Saving miners who are now trapped underground would be a rescue, but upgrading pit machinery to reduce the risk that accidents will happen in the future would be beneficence, but not rescue. As Chapter 7.2 discussed, the chief ethical debate in this area is if the claims of those now in peril really are more pressing than those of unidentifiable individuals who may get into peril at some point in the indeterminate future.

While some, such as Singer (1972), argue that obligations of beneficence are just as stringent as those of rescue, they do so on the basis of a moral argument, rather than—as Emerson appears to do—simply re-categorizing a case of beneficence as one of rescue. If we followed Emerson's usage, and allowed reducing risk to unidentifiable people over the long term to count as rescuing them, then it seems that more or less anything would count as a rescue. Most ordinary public health activity, such as routine immunization, or health and safety inspections of restaurants, would count as rescuing those unidentifiable individuals who would then not contract disease.

It would seem better to acknowledge that the eradication campaign does not *rescue* the people who do not get polio in the future. Rather, it permanently removes a health risk of a certain kind from their environment, and so makes it the case that no one will in the future have to be rescued from this health risk. This is an important benefit, but it is not one that makes the health benefits of eradication different in kind from those of other public health policies. As the next section explores, this is the ground for a more successful argument in favour of eradication policies.

10.5.4   Eradication as Ordinary Health Policy

Malaria creates a great burden of disease—causing more than 400,000 deaths per year, and over 200 million cases (World Health Organization 2018a). If an effective vaccine became available, and a successful eradication campaign then reduced to zero the burden of disease from malaria for the remainder of human existence, this would provide an extraordinarily large health benefit (Liu et al. 2013). While there is no special reason to opt for eradication policies as such, eradicating disease is clearly one way of meeting more general desiderata of public health policy—reducing the burden of disease equitably and efficiently. Eradication policies will sometimes have a more favourable balance of burdens and benefits than other competing health interventions—and in such cases they should be chosen.

Standard cost-effectiveness tools struggle to accurately account for the benefits of ordinary national vaccination campaigns (Beutels et al. 2008). Accounting for the benefits of eradication campaigns is significantly more difficult. In what follows, I shall aim to sketch some of these additional problems, and argue that they should not stand in the way of eradication campaigns.

The first difficulty relates to uncertainty. It is extremely difficult to globally eradicate a disease. Only one such attempt has so far succeeded in humans, so it would be unrealistic to think that any given eradication campaign could be guaranteed success. Where an eradication campaign fails, it can fail more or less gracefully. It can fail gracefully where, despite not leading to global eradication of a disease, it leads to a significant and sustained reduction in prevalence of the disease, or it can fail less gracefully, leaving no sustained reduction in the prevalence of the disease, and a trail of negative associations that makes it more difficult to mount eradication campaigns in the future. Constructing a model for the prospective cost-effectiveness of eradication campaigns is thus very challenging, though progress is being made here (Barrett 2013).

Second, there are both ethical and cost-effectiveness reasons for thinking that eradication campaigns should aim to go big and go fast (Omer et al. 2013). If an eradication campaign lingers around for a long time in the 'last mile', the cost for preventing each additional case will go up exponentially. The same precautions still need to be taken, and all the surveillance, but the number of people who are actually suffering from the disease will be very small. It is at this point that vaccine hesitancy is more likely to become a problem—as individuals may not unreasonably question whether they themselves stand to benefit from the vaccination, or if protection against a vanishingly rare threat is worth having. If policy-makers take the view that eradication should continue to be pursued only where the cost for each individual case remains within tolerable bounds, then they are likely to give up before the job has finished—meaning that there will be continued flare-ups of the disease, with the net result that the disease will never be eradicated (Thompson and Tebbens 2007).

Third, and most difficult, there is a deep question about how to weigh even *successful* eradication campaigns in the balance against other uses of healthcare resources. Disease eradication brings its true benefits only over the long term, while healthcare spending tends to focus on short- to medium-term benefits. If we assume that it is equally as important to save a life in fifty or one hundred years' time as it is to save one now, then it would seem that we should devote a very great proportion of our current healthcare resources to eradication campaigns. As Murray put this point in setting out the initial framework for the Global Burden of disease report:

> if health benefits are not discounted, then we may conclude that 100% of resources should be invested in any disease eradication plans with finite costs as this will eliminate infinite streams of DALYs [disability-adjusted life years] which will outweigh all other health investments that do not result in eradication.
>
> (Murray 1994: 440)

Murray drew the conclusion that in order to avoid this paradox, future health benefits should be subject to a discount rate. This conclusion seems surprising: if the expected total health benefits of eradicating a disease such as malaria really were vastly greater than, say, improving control of diabetes, wouldn't this be a strong argument in favour of eradication?[20]

While the terrain here is complex, we argued in Chapter 7.6 that there seems to be no good reason to apply large discount rates to future health benefits, even if there are good reasons for significantly discounting other future goods. There seems to be no reason to think that the mere fact that suffering or death is proximal in time provides a reason to think it more ethically important, any more than there is a reason to think that suffering or death is proximal in space does.

There are also some epistemic questions that could be raised, quite separately from claims about any intrinsic lesser importance of lives in the future. For example, there is a chance that the disease will no longer be a problem in the future and if so, it would be a waste of resources to spend money now on eradicating it, rather than treating another health condition. Perhaps there will not be any human beings around to still gain the benefit of the disease's being eradicated—in which case expending the time and effort now to complete the last mile of the disease's eradication would turn out to have been futile. However, it seems implausible to think that any such discount rate based on the possibility of unforeseen disaster should be large, as 'even a 1% discount rate implies that there is a 50% chance that the world will end in 69.7 years' (Anand and Hanson 1997: 695).

An example that Parfit (1984: 453–7) initially devised in a different context provides a useful way of conceptualizing what is at stake here. Suppose we are thinking about three scenarios for the future of malaria.

1. Status quo.
2. A malaria control campaign reduces the current burden of malaria by 99 per cent.
3. An eradication campaign globally eradicates malaria.[21]

It is obvious that, other things being equal, 3 is better than 2, and 2 is better than 1. But how much better is the successful eradication campaign than the control campaign, which merely reduces the burden of its disease to 1 per cent of its current level? Many people would assume that the successful eradication campaign is only marginally better than the successful control measures. But this is to ignore the fact that if we simply reduce the current burden of malaria by 99 per cent,

---

[20] Later editions of the Global Burden of Disease report abandoned discounting of future health benefits.

[21] Parfit's original example concerns a choice between (1) status quo, (2) a nuclear war that destroys 99 per cent of the existing population, and (3) a nuclear war that destroys all the existing population, but the underlying point is the same.

then malaria will (absent some further attempt at eradication, or dramatic change to the environment) continue to cause illness and death for the rest of human history. The likely total benefits of a successful eradication campaign are thus huge in comparison to the control campaign.

Some commonly advanced arguments for thinking that eradication is an ethically exceptional goal are weak. But my aim has not been to oppose eradication as a policy goal, but to give a better explanation of why it is compelling. The main reason for advocating eradication (in cases where it is feasible to do so) is none other than the future health benefits that it provides. There is no good reason to discount future health benefits for reasons other than those of uncertainty; and discounts as a result of uncertainty should be relatively small. And once we recognize this, then the sheer scale of the health benefits that eradication offers gives us a good reason to attempt it in cases where it is judged feasible.

10.6 Mandating Vaccination

We have seen how the combination of the significant overall benefit and protection of the vulnerable provided by vaccination programmes—whether they aim at disease control or at eradication—is sufficient to generate a prima facie ethical obligation to agree to be vaccinated, and that in cases of ordinary vaccination programmes, the vaccination will also be in the *ex ante* interests of those vaccinated. It is notable, however, that in many cases even where vaccination is convenient and costless, states may struggle to gain sufficient uptake for herd immunity, let alone complete eradication of a disease.

Much ethical debate on vaccination has thus centred on the justifiability of mandating vaccination, and hence of using the force of the state to ensure vaccination—either by measures such as fines, or not allowing children to attend publicly run schools if they are not up to date with their vaccinations. As we saw in Chapter 6, public health measures that interfere with liberty are not at all unusual, and include labelling of poisons, routine health and safety inspections of restaurants, product safety standards, and speed restrictions on roads. However, the fact that vaccination is a medical intervention raises a question about whether it should be framed in the light of commonly assumed rights such as that to bodily integrity and the right to refuse medical treatment.

Whether government action to promote and protect health is ethically justifiable depends crucially on whether the interferences with individuals' lives are necessary and proportionate, as well as the benefits to be achieved. Chapter 5 advanced four principles to help reconcile the goal of protecting and promoting health with that of protecting autonomy and liberty. These provide a good start for thinking about the justifiability of policies for the maintenance of herd immunity that go beyond exhortation in the direction of mandating vaccination. We can specify these principles for the case of vaccination policy as follows:

1. The size of benefit to be gained or size of harm to be avoided by a given vaccination policy matters. Other things being equal, the greater the expected benefit and the greater the expected harm to be avoided, the stronger the argument in favour of an intervention.
2. The extent to which the population endorses or consents to the policy matters. Other things being equal, the greater percentage of the affected population who endorse an intervention (and the more enthusiastically they do so) the stronger the reason in favour of the policy.
3. The ability to make autonomous choices matters. Other things being equal, the more significant a choice is, the more important it is that a person has the opportunity to make a genuine or authentic choice and the more problematic it is to interfere with their choice.
4. Liberty matters. Other things being equal, the more coercive a policy is, the more problematic it is.

Whether a policy is necessary and proportionate depends on whether the ends of the policy could be achievable with an approach that involves less interference with liberty and with autonomous choices.[22] This is not a fact that is fixed separately from the way that the policy and the actions of the government more broadly are interpreted by its citizens. In some circumstances, making vaccination mandatory may *increase* public confidence, if it is interpreted as being enacted by the government *because* the dangers of not being vaccinated are sufficiently high, and because of the reasonableness of asking everyone to play a role in protecting the common good.

However, in other contexts, mandating vaccination may exacerbate vaccine hesitancy, and make it more difficult to maintain herd immunity. Mandating a policy that was previously supported on a voluntary basis may contribute to

---

[22] As the onset of the Covid-19 pandemic made clear, in an emergency context in which cases of disease are doubling every few days and a health system threatens to become swamped, it will usually be better to act with sufficient force to ensure that the outbreak is brought under control—even if this involves the temporary imposition of measures that might otherwise look excessive. Much of the reason for going hard and early is the sheer difficulty in an early pandemic of monitoring whether a containment policy has been effective, and shifting quickly enough to a more coercive one if it is necessary to keep the disease under control. In ordinary contexts, it is usually feasible and indeed desirable make an intervention, and see if it is sufficient before moving to a more coercive or costly intervention. In the early stages of a pandemic this is not possible. Data will be patchy, and will lag behind the current state of infections (e.g. because of incubation periods). Any pandemic policy that aims to begin with the least infringement of liberty and increase stringency as necessary will create an obvious risk that politicians will end up putting in place an escalating series of interventions, each of which might have been sufficient to control the epidemic at an earlier stage, but which together fail to bring the disease under control, culminating in a need for very stringent lockdown measures over a much longer period of time than would have been necessary if decisive action had been taken early on.

undermining public confidence in the policy (Frey and Jegen 2001). The legal enforcement of smallpox vaccination in 1854 in England provides a classic example of such unanticipated consequences. The enforcement backfired initially, resulting in a decrease of uptake and increase in smallpox-related mortality. While uptake rates did improve over time, and other European countries with compulsory smallpox immunization had lower associated mortality rates by the 1870s (Winslow 1903), an unintended consequence of the legal enforcement of smallpox vaccination in England was the birth of the anti-vaccination movement in England which was influential in the nineteenth and start of the twentieth centuries (Wolfe and Sharp 2002).

Overall, policymakers must perform a delicate balancing act: if vaccine hesitancy is widespread, this may be perceived to make policies of mandatory vaccination necessary; but the fact of widespread vaccine hesitancy may itself undermine the perceived political legitimacy of so doing. Which vaccination policies are justifiable in a given case will depend (among other things) on the level of general consent to the policy, and the significance of the choices interfered with. The level of consent will obviously vary relative to culture and time; and we will also have to take account of local differences in which choices are believed by particular communities to be significant.[23]

The touchstone of the right to public health account is justifiability of public health activity *to* individuals. So any policy of mandatory vaccination needs to take seriously the need to justify to individuals who are coerced. Given that enforced vaccination will infringe on the bodily integrity of individuals and will be strongly resisted by some for reasons of religion or personal belief, this is a fairly high bar.[24] So it is reasonable to think that, as Verweij and Dawson (2004: 3123) put it, participation in vaccination programmes 'should, generally, be voluntary unless compulsory vaccination is essential to prevent a concrete and serious harm'. Realization that vaccination programmes require a high level of buy-in

---

[23] See Chantler et al. (2019) for some case studies of different approaches to mandatory vaccination.

[24] Mandatory vaccination policies also need to take a stance on exemptions. Navin and Largent (2017) helpfully distinguish three types of strategies for managing objections: Eliminationism (not allowing non-medical exemptions), Prioritizing Religion (allowing only religious-based exemptions, and not ones based on other personal beliefs), and Inconvenience (allowing both religious- and personal belief-based exemptions, but making it inconvenient enough to receive such an exemption that those whose objection is not strongly held are likely to be deterred).

Where a policy is one to which someone has a genuine objection of conscience, it is a serious matter to mandate overriding this, even if a position to which someone has a conscientious attachment seems wrong-headed from the perspective of the mainstream. The right to public health framework would thus suggest that where there are mandatory vaccination policies, there will usually be good reasons to allow some non-medical exemptions. It is difficult to articulate what makes an objection a religious one, and within the context of a secular state it is also difficult to justify why religious objections should be afforded a special status. So, it is difficult to justify Prioritizing Religion. To the extent that there are genuine worries about people gaming the system by claiming exemptions under false pretences, it would be fairer to keep a cap on numbers in a way that is neutral between religious and nonreligious reasons, perhaps by making exemption less convenient.

from citizens, and that mandatory vaccination may decrease rather than increase buy-in should lead policymakers to think systemically about how to build and maintain trust in vaccination.[25]

10.7 Conclusion

This chapter has examined communicable disease through the lens of an ecological approach. It analysed three different problems: drug resistance, ensuring herd immunity, and disease eradication. In each case, the solutions suggested required understanding the dynamics of communicable disease in complex systems terms.

Each discussion aimed to avoid getting trapped by too narrow a view. Whereas some have focused on antimicrobial resistance as a problem of the distribution of fixed scarce resources, we saw that such questions are downstream of more fundamental ones about how to minimize the occurrence of infections for which antimicrobials are required. While there are some bad arguments for why eradication should be pursued, seeing eradication for what it is—an extraordinarily effective way of improving the lives of people far into the future—allows us to understand better the ethical imperative to do so, and the ethical questions about what sacrifices now are appropriate to ensure this. Whereas much philosophical debate has focused on the justifiability of mandatory vaccination, we saw that the deeper problem is how to maintain public trust in a system in such a way that coercion is not required.

Health systems have goals that are varied enough that it is difficult to imagine that there could be a single number that could sum up what they do and which they could be more or less efficient at. Even if we artificially restrict our focus to reducing the biomedical disease burden of a particular infectious disease, there are a number of goals, differing on at least three dimensions: space, time, and risk concentration. What is efficient relative to one end may not be efficient relative to others; and these complexities are significantly exacerbated by the phenomenon of drug resistance.

What counts as a fair use of antibiotic effectiveness, and whether an eradication programme is worth attempting, will depend quite thoroughly on the geographical space and the time period under consideration. At one extreme would be the goal of successfully treating one patient here who is ill now—a goal that is maximally confined on each of the three dimensions of space, time, and risk concentration. At the other extreme would be a global eradication campaign, where

---

[25] Other areas of policy also face similar problems of trust and social licence—in particular, usage of healthcare data for purposes beyond a patient's direct care. For more on such topics, see Taylor and Wilson (2019).

the goal is that no one in any part of the world will either suffer the disease or even be subject to the risk of it now or in the future.

These questions about territory, timescale, and prevention versus treatment arise for treatments for noncommunicable diseases, and were considered at length in Chapter 7, but they are sharpened by the consideration of communicable diseases, and sharpened again when it comes to drug resistant infections. Given the movement of people across the globe, outbreaks of communicable disease in one country can threaten other countries. Drug resistant infections considerably raise the stakes, because it means that not only will there be additional cases of the disease, but that treatments that would have been successful against these new cases will now no longer be so.

Deciding where in this three-dimensional cube of time, space, and risk concentration the priorities for a particular health system lie will make a difference to what counts as optimal. Questions that policymakers in their particular contexts will need to answer include: what weight should be given to prevention versus treatment; whether vaccines should be targeted at those at high risk of mortality from a disease, or should be focused on those most likely to act as vectors, or alternatively given to the whole population; and whether there are instrumental or intrinsic reasons to deploy resources to help control infectious diseases outside of their territory; and whether success should be judged from the perspective of the short term, the medium term or the long term. I have not attempted to provide a definitive answer to where any particular health system should place its priorities in this three-dimensional space, but hope to have provided citizens and policymakers with a set of useful tools for developing their own answers to these questions through public deliberation.

11
Conclusion

11.1 Introduction

This book has pursued both practical and theoretical aims. At a practical level, it has aimed to provide a source of advice and help for citizens and for policymakers in different roles about some foundational questions that arise for responsible policymaking. These have included epistemic questions about the ways in which evidence is relevant for and to policymaking; why evidence can be of high quality in one context, but yet not useful in other contexts; and why policy interventions that incorporate a lucid understanding of the complex interactions in a particular domain are less likely to be thwarted by policy resistance.

The book has provided advice on how to specify and reconcile the values that should inform policymaking; how to think about the relationship between the individual and the state; and why it is important to be able to justify all public policies to citizens as individuals. Recognizing the importance of context, it has worked through some more specific examples of how core values such as responsibility and health equity should be operationalized.

Public policy needs ethical and philosophical reflection in order for the choices made by policymakers to be defensible, let alone wise; but not just any form of ethical and philosophical reflection will do. So at a theoretical level, the book has been a sustained inquiry into how to do philosophy in a way that is helpful for public policy, and some of the ways that philosophy as a discipline needs to change to accommodate this.

Philosophers need to shift away from implicitly viewing policymakers as outside of the system that they are attempting to steer, towards a dynamic view that sees policymakers as actors within the same system, and who will themselves be acted upon. While philosophy is thus dethroned from a position of offering blueprints or peremptory demands to policymakers, it does still have a distinctive role to play in clarifying the epistemology of public policy, thinking through the implications of external validity and of performativity, specifying how to think about the values that should guide policy, and providing ethical frameworks for policymakers to build on.

11.2 Lessons About the Epistemology of Public Policy

The crucial epistemic question for public policy is how best to move between evidence—in whatever form, whether randomized control trials, or opinion surveys, or mathematical models—and the policy decisions that need to be made. Given that policymakers are always dealing with complex systems, there is reason for them to be sceptical that there will be an accurate mapping between the outputs of models that are provided to them, and what will actually happen.

Nonetheless, even simple models can be invaluable in understanding the contours and dynamics of a problem: for example, examining the effects of changing different model parameters and their interconnections may provide researchers and policymakers with a crucial set of insights for scenario planning. Thus, a model can help policymakers to understand the kinds of likely impacts that the closure of a hospital might have on access to care, even though the flows of actual patients are unlikely to be accurately and precisely predicted by the model.

Simple models may also be useful if they allow researchers intelligently to question assumptions that previously seemed obviously correct. For example, it may be natural to think that if the measured population of a particular animal varies erratically over time, then this would indicate either some kind of measurement error, or significant changes in the environment but as May (1976) shows, a simple and deterministic model of animal populations can also lead to such apparently random fluctuations.

There is no problem in taking simplified models and using them for exploring the dynamics of real-world problems, if this is done judiciously. Indeed, it is hard to make progress in working out how to intervene in complex systems without doing this, but it is crucial to be clear about what the relationship is between the model and the real-world problem. Rigour requires apportioning confidence reasonably given the state of the evidence and the models. This requires understanding the quality and sparsity of the available evidence, and the choices that are made in collecting it. It requires also understanding the choices made in going from the 'raw' data to a processed model of this data, and in moving from one or more models of the available data to an explanatory model that will be used to help determine potential sites for intervention. Each of these stages requires choices, and often each of the choices could reasonably have been taken differently.

Taking complex systems seriously implies that questions that are often thought about under the domain of scientific method or philosophy of science—about what and how to measure, model design, and so on—need to be integrated into the approach taken in ethics and political philosophy. But it is also the case that doing science well requires making appropriate ethical judgements in model construction, and in the use of models. Which variables matter most and which should be measured when it comes to public policy is not a question that can be decided without ethical reflection. Chapter 9 discussed at length how it is unhelpfully vague to make claims about the unfairness of health inequalities without specifying *which* inequalities, how we are to measure them, and what would count as an improvement. Doing so requires entering into trade-offs about the costs of deriving different datasets, and the quality of available data, as well as judgements about when it is useful to concentrate for policy purposes on univariate, bivariate, or higher dimensional disparities in variables.

Wherever there is a move between evidence or models and something to be done on the basis of this evidence, the problem of external validity will need to be addressed. There will nearly always also be a trade-off between avoiding type I errors ('false positives'), and type II errors ('false negatives'). This is not just an epistemic trade-off, but rather also an ethical trade-off about how important it is to avoid detecting something that is not there, as opposed to failing to detect something that is there in a particular context (Rudner 1953).

To give just one example, many interventions that are proposed in precision medicine amount to secondary prevention on the basis of big data and genomic screening, and bring with them significant risks of overdiagnosis and overtreatment (Vogt et al. 2019). While the idea of a precise intervention sounds superficially much more attractive than an untargeted one, the crucial question is the overall balance of harms and benefits that it would lead to. There are often strong reasons for health systems to prioritize untargeted primary prevention measures that reduce population health risks, over highly targeted interventions which aim to reduce the effects of risk factors on citizens' health through pharmaceutical intervention, while leaving the sources of population-level risks unchanged (Wilson 2021).

11.3 External Validity, Performativity, and Philosophical Methodology

Some, and perhaps many, philosophers' ideas about methodology in ethics, and in particular what makes for rigour in ethical thinking, are deeply mistaken. These philosophers have too easily assumed that clever manipulation of simple and unrealistic thought experiments will give insights into ethical principles that are timeless and universal, and that these timeless and universal principles in turn can helpfully inform decision-making in policy. Although they may not always say so explicitly, such philosophers presuppose that the flow of changeable patterns of events, lives, institutions, and norms is somehow misleading as to the true nature of things, and that at a deep level, ethical reality is simple, static, and timeless.

This book has argued both explicitly and implicitly for a very different conception of philosophy. On this view, philosophy is at heart a form of engaged problem-solving. Philosophical thinking becomes necessary largely because of the ways in which social, political, and technological change disrupts patterns of activity that have become habitual. What has previously been able to form part of an unchallenged background of common activity becomes newly problematic. For example, the invention and spread of writing as a medium profoundly changed the nature of story-telling, of education, and the ways in which know-ledge could be shared—something which Plato explored in his dialogue the *Phaedrus*. Much more recently, the rise of big data, and the concomitant ability to infer far more about individuals than was previously possible, have raised questions about both the nature of privacy and what may legitimately be inferred from infor-mation that has been made public (Nissenbaum 2009; Rumbold and Wilson 2019).

The approach to philosophy that I have advocated takes change, whether of things, or persons, or institutions, or norms, as what is fundamental and explains regularities, patterns, and fixed points in these flows in terms of the interplay of dynamic processes that maintain these variables within narrow boundaries. Much of what seems solid and static has evolved as a result of dynamic processes, and remains apparently static because of the interplay of dynamic processes. Order and regularity often arise bottom-up from the effects of different elements of sys-tems regulating and affecting one another, rather than top-down because it is necessary that things be so.

This was not a perspective that I brought with me when I first started thinking about public health ethics in 2007, but one that I found was called for by the nature of the problems themselves. In particular, the kind of challenge posed by thinking about drug resistant infections—in which ethical reasoning needs to be able to theorize how to respond to threats to health that alter dynamically across space and time—seemed from the beginning difficult to specify adequately using tools which presuppose that ethical reality is unchanging. The more I continued on the project, the more it came to seem that drug resistant infections were not an exception that needed somehow to be accommodated within ethical frameworks designed for static problems, but that they were in fact the usual form of ethical problems in public policy. Thus, what was required was to find a framework that would be suitable for thinking about drug resistant infections, and then think how this framework could be generalized to consider other problems that arise in public policy.

As I have attempted to draw the reader's attention to from time to time, perspectives that give precedence to the static and unchanging are difficult to reconcile with the fundamentals of geophysical, biological, and social science, notwithstanding their very respectable pedigree within philosophy, and their congruence with the ways that many feel about their place in the world. Geophysical features which common-sense experience may take for granted as unchanging, and which earlier humans used to ground their other measures (such as the length of a day, the number of days in a year, the location of the different landmasses on the surface of the earth, and the chemical composition of the atmosphere and the oceans) have all changed greatly over geological timescales and continue to change.

These changes have, to a significant degree, been driven by the dynamics of life itself—for example, the early earth atmosphere and oceans contained only trace amounts of oxygen, and it was only after the Great Oxidation Event that aerobic organisms became feasible; and only after considerably more accumulation of oxygen that the necessary energy requirements could be met for animals that moved around rapidly (Judson 2017). The coming of the Anthropocene has drawn further attention to how profoundly human activity has influenced planetary ecosystems (Wilson 2018). Basic social forms and structures, such as agriculture, writing, law, property, money, science, mathematics, kinship, ethical norms, and punishment, each also have a history, have involved an unfolding sequence of innovations and adaptations, and continue to do so.

One main challenge in popularizing a new way of ethical thinking is the old adage 'if it ain't broke, why fix it?' That is to say, where philosophers have been using a set of tools for thinking for many years, and believe that these tools have served them well, they are unlikely to give them up solely because someone points to a kind of problem—such as drug resistant infections or mental health stigma— for which the existing techniques are poorly calibrated. The temptation is instead to work to diminish the importance of the new kind of case and to keep focus on the cases that the old approach *can* handle; and if pressed to, find a way of translating the new kind of case into the existing framework.

The book has attempted to address this worry head on, by asking what the world would have to be like *if* thought experiment-led approaches that aim to uncover a static ethical reality were to be a rigorous and useful approach to doing ethics. The first point in this journey was a set of observations from Dennett (2006) and Kitcher (2011c), who between them argued that there is a vast and probably infinite range of things that can be set up and debated as philosophical problems—and in each case a literature could be built with increasingly subtle distinctions, arguments, and counterarguments. So the mere fact that there is a niche that is susceptible to increasingly refined debates using some recognizably philosophical techniques does not by itself show that this activity is useful, or should be a priority for the distribution of scarce research grant funding.

How to prioritize philosophical problems for reflection and which methods to use in this reflection is a crucial question. I have attempted to make some progress by characterizing the purpose of ethics, and then relating the usefulness of philosophical methodologies to their success in getting us closer to this goal. On the view I have defended, ethical reflection and debate is necessary because of the ethical disagreements, problems, and dilemmas that arise for agents in the course of living our lives, including in the making and implementation of public policy. Ethics begins in problems that occur in living together, and there are thus reasons why it must remain accountable to ethical practice, if its debates are to avoid becoming sterile and being overtaken by the display of academic virtuosity for its own sake.

One important implication of taking practice seriously within ethics is that methods for improving ethical judgements which rely heavily on abstraction and simplification face a double challenge of translation. First, the ethically relevant features of a situation that raises an ethical problem need to be translated into a simplified and abstract form for philosophical analysis. Second, once the thinking has been done on the simplified and abstract analogue of the ethical problem, these results then need to be translated back into a form that can shape wise responses to the real-world ethical problem that called for ethical reflection.

Chapter 3 highlighted some of the ways in which information that is relevant for acting in the real world can get lost in the course of the double process of translation. In particular, thinking rigorously and elegantly about a toy problem does not imply that the results of this reflection will be useful for making wise decisions about messier and more complex real-world problems. Among other methodological challenges, different underlying causal structures may be presup- posed in the thought experiment from those in reality, and thought experiments may exclude normative contextual interactions that occur in real-world cases. These difficulties caused by two-way translation are challenging enough for thought experiment-led approaches to ethics, but still allow philosophers to maintain the idea that thought experiments can provide a useful though often unreliable source of insight into pre-existing moral truths, and that it is sometimes possible to move from these insights to wise actions in the real world.

Chapter 4 introduced performativity as a deeper challenge for various approaches to ethics that foreground static principles and abstraction. The challenge of performativity strikes not at the idea that abstraction and simplification can be a useful tool in ethical thinking, but at the idea that the ethical reality into which philosophers are trying to gain insights is static and discoverable, rather than changeable and contingent. Merton's (1948) parable of the run on the Millingville bank (Chapter 4.6) showed vividly how the way things turn out is often materially affected by people's beliefs and interpretations about what others will do.

Performativity affects not just facts about the way the world is, but the concepts by which we navigate ethical space, and the institutions through which societies aim to ensure structural justice. There is a wide range of cases in which ethical problems and their solutions exist only in relation to human cultural formations— and in fact such cases are the norm rather than the exception. The book has seen a steady accretion of concepts that are crucial for theorizing in public policy but which cannot be understood other than in performative terms, such as public trust, reasonable expectations, policy resistance, self-reinforcing policies, substantive responsibility, structural justice, race, gender, disability, stigma, and oppression.

These concepts do not affect public policy singly but in complex constellations. For example, where a novel public health policy seeks to combat intersectional inequalities that currently favour some powerful groups, and these groups have the weight of tradition behind them, then it is vital that consideration is given to how the policy can be enacted in a way that avoids policy resistance and builds rather than loses public trust over time.

Performativity does not defeat thought experiments and other forms of abstraction, but it does force a rethink of their purpose. Thought experiments as they are usually framed within philosophy turn on what a single actor does or should do, while everything else is held fixed. But by their nature, cases in which performativity matters turn on the combined effects of what a plurality of separate agents do, where each agent acts in anticipation of what others will do, and the effects of these actions and interactions iterate over time. This requires thinking through the different perspectives, values, and strategies that other agents may have, and gives any model of a performative scenario a dynamic and ecologically defined set of outputs, rather than a single solution.

11.4 Public Health, Ethical Frameworks, and the Need for Improvisation

This book has not just defended a methodology for how philosophy should seek to inform public policy, but also a substantive ethical framework based on ideas of equality, democratic accountability, and performativity. This framework emphasizes that one implication of equal citizenship is that public policy interventions must be justifiable *to* the individuals affected by them. The mere fact that a policy would create an aggregate benefit at a population level does not provide a sufficient justification for adopting it.

While policy interventions need to be justifiable *to* individuals, this does not imply that they should be targeted *at individuals*. Individualistic justification is not only compatible with, but will often require, interventions that are targeted upstream from individuals, and which have their effects only indirectly. Intervening at a broader social level to improve health will often be easier to justify to individuals than interfering with individuals' lives directly, given that it is easier for upstream interventions to bring risk reductions without disproportionate inconvenience than it is for interventions that require individuals to change their behaviour.

Promoting health in the narrow biomedical sense is only one of a number of legitimate goals for governments. Many individuals will also treat biomedical health as only one of a number of things that matter to them. There will be reasonable disagreement both amongst citizens and amongst policymakers about the appropriate weight to give to the protection and promotion of health. Giving 'too much' weight to health may threaten other priorities. Policymakers in public health will often need to seek synergies with other goals, rather than simply asserting the priority of public health.

Given these constraints, I have argued that the best way to articulate the broad shape of governments' duties in respect of public health is the idea of the right to public health. Asserting this right allowed us to explain why it is mistaken for governments to adopt a one-sided approach, in which they respond only to the ethical worry of creating an overprotective 'Nanny State'. It is equally if not more important for governments to avoid becoming a Neglectful State, which fails to take easy steps to reduce risks to population health, and as a result allows significant numbers to come to avoidable harm or death. Where a government (or other agents, such as commercial companies) fall short of taking the steps that are reasonably required to reduce population-level health risks, then it is appropriate to ask whether individuals' right to public health has been violated.

This high-level framework aims to clarify the ethical features of a range of choices that need to be made about goals and means in public health. In designing the framework, and thinking how it could be specified for the cases of substantive responsibility, health equity, and communicable disease, the aim is not to provide policymakers with a detailed set of instructions about how to organize public health policy. Rather, the aim is to provide a set of arguments, principles, and examples, which will need to be customized and further specified in line with a state's own traditions and priorities.

Policymaking is improvisation, and philosophers' role in public policy is to support policymakers' improvisation, not to supplant it. To say that policymaking is improvisation is not to say that policymaking should be unplanned, or theoretically uninformed. Improvising well requires a depth of experience, knowledge, and skill, in addition to spontaneity. It is no more plausible to think of public officials improvising wise responses to complex policy challenges without experience, evidence, models, and ethical principles than it is to think of someone who does not know how to play the instrument, and knows nothing of the history and repertoire of jazz, improvising a good saxophone solo. So my claim is not that preparation, modelling, theory, and evidence are unnecessary, but that they are not sufficient.

One central insight, which policymakers should incorporate in interpreting and building on the analysis of this book, is the need to integrate ethical analysis with rigorous analysis of systemic interactions. Realiszing that there may be different ways of sustainably 'solving' a problem that is shaped by performativity is the beginning of wisdom. We saw this in thinking about both responsibility, and vaccination policy. Whom to hold substantively accountable, and in what circumstances, is above all a question of systems design. It should not be treated as a question that can be answered once and for all, but one that will need to be rethought as contexts change. Similarly, what matters ultimately for vaccination policy is establishing effective control and reduction of disease. In some contexts this will be best served through establishing and maintaining public trust in a voluntary system, while in others it may require mandatory vaccination.

The message of this book is at heart a hopeful, though not necessarily an optimistic one. We make the social world; and the actions and inactions of individuals create and sustain the systems within which injustices and unnecessary suffering occur. We can, however, also remake worlds. Doing so is not straightforward; but this book will have provided you—whether you are a philosopher, a policymaker, or an interested citizen—with an array of tools with which to make a start.

Afterword

The first cases of sustained transmission of the SARS-CoV-2 virus, which causes Coronavirus disease 2019 (Covid-19), occurred in Wuhan province in China in late 2019. The disease then spread rapidly, with confirmed cases in 19 countries by 30 January 2020, when the World Health Organization declared a Public Health Emergency of International Concern.

The pandemic posed the severest test to local, national, and global health institutions. The requirement to contain a highly infectious virus that is spread by droplet transmission and aerosols, and for which no vaccine was available, saw many states introduce a range of very stringent public health measures. Interventions that would previously have seemed unthinkable, such as lockdown policies that penalized meeting friends or even leaving the house, were widely adopted. These measures, when allied with effective testing, contact tracing, and isolation of suspected cases, very significantly reduced cases, but they completely suppressed the virus in only a few countries.

As I was finalizing this work at the end of December 2020, 81 million Covid-19 cases had been confirmed, and 1.8 million deaths globally. Around 700,000 new cases were being reported per day globally, which was the highest since the pandemic had begun. A number of vaccines were in the early stages of rollout in a handful of countries, but it was still far from clear whether the first wave of vaccines would prove sufficiently effective to suppress the virus, how long immunity would last, and when such supplies would be widely available in resource-poor settings (Dyer 2020).

This book's draft manuscript was already with the publishers before the pandemic struck. In the months following the outbreak, I was involved in a number of projects that aimed to give ethical advice to policymakers in the UK. This included providing advice on the ethics of intensive care unit (ICU) prioritization to regional and national committees; advising the National Data Guardian for Health and Social Care on a range of Covid-related changes to processes for accessing NHS patient data for planning and research; membership of the Ethics Advisory Group for the NHS Covid-19 App, and advising on ethical principles for the distribution of vaccines (Campos-Matos et al. 2021). Much of this advice relied on the framework developed in this book, including Chapter 4's complex systems approach to public policy, Chapter 7's analysis of prioritization, Chapter 9's complex systems account of health equity, and, of course, the analysis of communicable disease in Chapter 10.

Despite having been deeply involved with providing ethical advice on Covid-19 during much of 2020, it still seemed rather too soon to write at length about it in the body of the book. The book as a whole aims to provide a philosophy of public policy, a framework of values, and a method that is helpful for thinking about a wide range of problems in public health and public policy more broadly. To have radically changed this framework in a short period of time to match the very unusual circumstances of the pandemic could have led to a significant risk of what modellers describe as 'over-fitting'—adjusting a model too much to respond to a small number of cases, and thus making it less useful for dealing with novel cases that had not previously been anticipated.

A major reason to worry about 'over-fitting' is that it is very difficult, if not impossible, properly to grasp the ethical implications of a global and fast moving public health emergency and bring it to into focus while it is still unfolding. Given the lead-in times imposed by book publishing, detailed and specific claims about Covid-19 risked a short shelf life, and possibly already being out of date by the date of publication. Even within the first few months of the pandemic, it had become apparent that as the facts changed, so did the ethical problems that it seemed were most crucial to address.

The question of prioritization of access to ICU, and whether it would be ethically permissible to remove a patient from a ventilator in order to accommodate other patients with a better chance of survival, seemed to be one of the most important ethical questions in the first months of the pandemic. It led to a flurry of papers and statements of ethical principles (White and Lo 2020). However, within months it became apparent that it was not only possible but clinically desirable to make a more sparing use of invasive ventilation in ICU, and much greater use of non-invasive respiratory interventions outside of ICU via cheaper and less labour-intensive techniques of continuous positive airway pressure (CPAP) and high flow oxygen (HFO) (Steiner 2020). Ventilator allocation came to seem less ethically pressing.

One of the chief messages of this book is that doing public policy well requires a high degree of contextual and systemic understanding. Even if a causal mechanism or an intervention works in one context, it should not be assumed that it will automatically work in another context. Even if an ethical principle encapsulates what should be done in a thought experiment, or even in a real-world situation, it should not be presumed that it will automatically be a good basis for wise action in another context. While this book's analysis has been much richer and incorporated many more perspectives than a thought experiment, its lessons also should not be assumed automatically to provide a basis for wise ethical decision-making in radically changed circumstances.

Thus, as the pandemic began to unfold, I was keenly aware that there were no guarantees that the book's analysis would continue to prove fruitful in radically altered circumstances. In the event, my framework's emphasis on systemic interconnections, performativity, and the need for structural justice proved just as useful in a pandemic as it was in prior conditions.

Chapter 10 examines the mathematics of communicable disease, making clear how critical it is to keep down the effective reproductive rate of the disease (R) so that it remains below 1. As the pandemic took hold, infectious disease modelling, and the effects on R of different interventions, became a main staple of the news. In early March 2020, cases were doubling every three days in the UK. Non-pharmaceutical interventions that reduced transmission, while still allowing cases to double more slowly, would not prevent hospitals from being overwhelmed, nor prevent an unacceptably high level of deaths. The only really viable strategy was to keep R below 1, by testing, tracing, and isolating, and the adoption of broader non-pharmaceutical interventions until an effective vaccine became available in large enough quantities.

Fairness, and maintaining public trust, emerged as key challenges for policies to contain the disease. While the pandemic began with appeals to solidarity and the acknowledgement of a new and shared risk that affected all—rich and poor alike—it soon became apparent that the risks of exposure to the disease, and of getting severely ill and dying if one did get it, were not shared equally. Initially, those who had jobs that brought them into contact with significant numbers of others and could not be done remotely faced a difficult choice between continuing to work and accepting heightened risk, or losing pay. Those whose jobs could be done adequately remotely did not face this choice. The obvious fact that it was predominantly white-collar jobs done by the wealthier and better educated that could be easily moved online, added to the sense of unequal risk.

Structural features such as race and social class, which were already responsible for significant inequalities in healthy life expectancy, magnified Covid-19-related risks of severe morbidity and mortality. The obvious connection between existing structural inequalities, and heightened death rates from Covid-19, raised deep questions about fairness and health equity—both in response to Covid-19, and more broadly. It was no accident that the Black Lives Matter protests became a global phenomenon, rolling up various different responses to structural racism in different countries.

The pandemic also exacerbated concerns amongst communities who had pre-existing suspicions about interventionist public health. Measures such as mandatory wearing of face coverings in public places became politicized—often recapitulating in an angrier and less coherent way some of the objections to liberty-impacting public health measures that are discussed at length in Chapters 5 and 6. Objectors included many who accepted the need for public health measures, but wanted more weight to be given to protecting the economy. Those advocating for less stringent public health measures, however, often failed to take into account a basic feature of communicable disease, namely that where public health measures are only moderately effective, and allow R to remain above 1, cases will continue to rise exponentially. So less stringent strategies were unstable—tending to lead either to economically destructive cycles of lockdown and then release, or an ethically dubious acceptance of large numbers of deaths.

Behind the heated debates were an important set of questions about fair distribution and redistribution of risk, which can helpfully be explored with the tools introduced in Chapter 7. The fact that a public health measure will bring benefits at a population level is not itself sufficient to make it ethically acceptable; it is crucial also that it is justifiable to individuals—even those who are expected to be left worse off as a result of the measures.

Non-pharmaceutical interventions such as closure of schools, workplaces, and shops aim to protect health system integrity and population health, but they also bring very different profiles of risks and benefits to different groups. School closures are one intervention among others that can bring a significant reduction to the R rate. However, while the risks posed by Covid-19 itself were very small for children and teenagers, the risks to children introduced by school closures often were not. A shift to online education, in which many children lacked access to adequate computer technology or an internet connection to participate, can significantly set back some children's interests.

With the closure of businesses such as bars and restaurants, owners found that they were still liable for rent and other business expenses but without the ability to earn a living. Many jobs became temporarily non-viable, and some proprietors who had spent a long time building up profitable businesses went bankrupt. Those who benefited most from the reductions to risk of infection from Covid-19 introduced by such lockdown policies were those at significantly higher risk of severe morbidity or death from the disease—most of whom were over 65. Lockdown policies thus involved a large-scale redistribution of risk from the old to the young (John 2020).

Maintaining public trust while pursuing such policies required governments convincingly to make the case that these risk redistributions were fair. Governments needed first to mobilize solidarity between citizens—explaining why many should endure either a risk either to their livelihoods, or the annoyance of taking additional precautions, in order to protect others who were vulnerable to more serious harms. Those who were going to lose out, particularly those who were unable to work or to run their businesses, needed to be compensated and so make it not just possible, but easy, to play their part in supporting public health measures. Some governments did this more effectively than others—and some reasons for this can be found in Chapter 8's discussion of responsibility, and others in the analysis of performativity and public trust in Chapter 10.

References

Ackoff, Russell L. 1979. 'The Future of Operational Research Is Past'. *Journal of the Operational Research Society* 30: 93–104.

Afshin, Ashkan, Mohammad Forouzanfar, Marissa Reitsma, Patrick Sur, Kara Estep, Alex Lee, Laurie Marczak, et al. 2017. 'Health Effects of Overweight and Obesity in 195 Countries over 25 Years'. *New England Journal of Medicine* 377 (1): 13–27.

Ahola-Launonen, Johanna. 2018. 'Hijacking Responsibility: Philosophical Studies on Health Distribution'. PhD Thesis, Helsinki University. Available from: https://helda.helsinki.fi/bitstream/handle/10138/241213/HIJACKIN.pdf?sequence=1.

Albertsen, Andreas, and Carl Knight. 2015. 'A Framework for Luck Egalitarianism in Health and Healthcare'. *Journal of Medical Ethics* 41 (2): 165–9.

Alexandrova, Anna. 2017. *A Philosophy for the Science of Well-Being*. Oxford: Oxford University Press.

Allotey, Pascale, Daniel Reidpath, Aka Kouamé, and Robert Cummins. 2003. 'The DALY, Context and the Determinants of the Severity of Disease: An Exploratory Comparison of Paraplegia in Australia and Cameroon'. *Social Science & Medicine* 57 (5): 949–58.

Anand, Sudhir, and Kara Hanson. 1997. 'Disability-Adjusted Life Years: A Critical Review'. *Journal of Health Economics* 16 (6): 685–702.

Anderson, Elizabeth. 1993. *Value in Ethics and Economics*. Cambridge, MA: Harvard University Press.

Anderson, Elizabeth. 1999. 'What Is the Point of Equality?' *Ethics* 109: 287–337.

Anderson, Elizabeth. 2010. 'The Fundamental Disagreement Between Luck Egalitarians and Relational Egalitarians'. *Canadian Journal of Philosophy* 40: 1–23.

Anomaly, Jonathan. 2011. 'Public Health and Public Goods'. *Public Health Ethics* 4 (3): 251–9.

Anomaly, Jonathan. 2012. 'Is Obesity a Public Health Problem?' *Public Health Ethics* 5 (3): 216–21.

Anomaly, Jonathan. 2015. 'Public Goods and Government Action'. *Politics, Philosophy & Economics* 14 (2): 109–28.

Antonovics, Janis, Jessica L. Abbate, Christi Howell Baker, Douglas Daley, Michael E. Hood, Christina E. Jenkins, Louise J. Johnson, et al. 2007. 'Evolution by Any Other Name: Antibiotic Resistance and Avoidance of the e-Word'. *PLoS Biology* 5 (2): e30.

Arneson, Richard J. 1989. 'Equality and Equal Opportunity for Welfare'. *Philosophical Studies* 56 (1): 77–93.

Arneson, Richard J. 2018. 'Dworkin and Luck Egalitarianism: A Comparison'. In Serena Olsaretti, ed., *The Oxford Handbook of Distributive Justice*, 41–64. Oxford: Oxford University Press.

Aronowitz, Robert A. 2001. 'When Do Symptoms Become a Disease?' *Annals of Internal Medicine* 134: 803–8.

Arras, John. 2016. 'Theory and Bioethics'. In *The Stanford Encyclopedia of Philosophy*, edited by Edward N. Zalta, Summer 2016. Metaphysics Research Lab, Stanford University. Available from: http://plato.stanford.edu/archives/sum2016/entries/theory-bioethics/.

Arthur, W. Brian. 1989. 'Competing Technologies, Increasing Returns, and Lock-In by Historical Events'. *The Economic Journal* 99 (394): 116–31.

Arthur, W. Brian. 2014. *Complexity and the Economy*. Oxford: Oxford University Press.

Asada, Yukiko. 2013. 'A Summary Measure of Health Inequalities'. In Nir Eyal, Samia A. Hurst, Ole F. Norheim, and Dan Wikler, eds, *Inequalities in Health: Concepts, Measures, and Ethics*, 37–51. Oxford: Oxford University Press.

Asada, Yukiko, and Thomas Hedemann. 2002. 'A Problem with the Individual Approach in the WHO Health Inequality Measurement'. *International Journal for Equity in Health* 1 (1): 2.

Ashcroft, Richard. 2008. 'Fair Process and the Redundancy of Bioethics: A Polemic'. *Public Health Ethics* 1 (1): 3–9.

Ashford, Elizabeth. 2003. 'The Demandingness of Scanlon's Contractualism'. *Ethics* 113 (2): 273–302.

Badano, Gabriele. 2016. 'Still Special, Despite Everything: A Liberal Defense of the Value of Healthcare in the Face of the Social Determinants of Health'. *Social Theory and Practice* 42(1): 183–4.

Balconi, Margherita, Stefano Brusoni, and Luigi Orsenigo. 2010. 'In Defence of the Linear Model: An Essay'. *Research Policy* 39 (1): 1–13.

Balshem, Howard, Mark Helfand, Holger J. Schüünemann, Andrew D. Oxman, Regina Kunz, Jan Brozek, Gunn E. Vist, et al. 2011. 'GRADE Guidelines: 3. Rating the Quality of Evidence'. *Journal of Clinical Epidemiology* 64 (4): 401–6.

Banerjee, A. V., and E. Duflo. 2011. *Poor Economics: A Radical Rethinking of the Way to Fight Global Poverty*. New York: PublicAffairs.

Baquero, F., T. M. Coque, and F. de la Cruz. 2011. 'Ecology and Evolution as Targets: The Need for Novel Eco-Evo Drugs and Strategies to Fight Antibiotic Resistance'. *Antimicrobial Agents and Chemotherapy* 55 (8): 3649–60.

Barrett, Scott. 2007. *Why Cooperate?: The Incentive to Supply Global Public Goods*. New York: Oxford University Press.

Barrett, Scott. 2013. 'Economic Considerations for the Eradication Endgame'. *Philosophical Transactions of the Royal Society B: Biological Sciences* 368 (1623): 20120149.

Barry, Nicholas. 2006. 'Defending Luck Egalitarianism'. *Journal of Applied Philosophy* 23 (1): 89–107.

Bateman-House, Alison, Ronald Bayer, James Colgrove, Amy L. Fairchild, and Caitlin E. McMahon. 2017. 'Free to Consume? Anti-Paternalism and the Politics of New York City's Soda Cap Saga'. *Public Health Ethics* 11 (1): 45–53.

Battin, Margaret P., Leslie P. Francis, Jay A. Jacobson, and Charles B. Smith. 2008. *The Patient as Victim and Vector: Ethics and Infectious Disease*. New York: Oxford University Press.

Bauman, Christopher W., A. Peter McGraw, Daniel M. Bartels, and Caleb Warren. 2014. 'Revisiting External Validity: Concerns about Trolley Problems and Other Sacrificial Dilemmas in Moral Psychology'. *Social and Personality Psychology Compass* 8 (9): 536–54.

Bayer, R., and J. D. Moreno. 1986. 'Health Promotion: Ethical and Social Dilemmas of Government Policy'. *Health Affairs* 5 (2): 72–85.

Bedau, Mark A. 1997. 'Weak Emergence'. *Noûs* 31: 375–99.

Beutels, Philippe, Paul Scuffham, and C. Raina MacIntyre. 2008. 'Funding of Drugs: Do Vaccines Warrant a Different Approach?' *The Lancet Infectious Diseases* 8 (11): 727–33.

Biggs, Michael. 2011. 'Self-Fulfilling Prophecies'. In Peter Bearman and Peter Hedström, eds, *The Oxford Handbook of Analytical Sociology*, 294–314. Oxford: Oxford University Press.

Billman, George E. 2020. 'Homeostasis: The Underappreciated and Far Too Often Ignored Central Organizing Principle of Physiology'. *Frontiers in Physiology* 11. Available from: https://www.frontiersin.org/articles/10.3389/fphys.2020.00200/full.

Birchley, Giles. 2016. 'Harm Is All You Need? Best Interests and Disputes about Parental Decision-Making'. *Journal of Medical Ethics* 42 (2): 111–15.

Blake, Michael, and Mathias Risse. 2008. 'Two Models of Equality and Responsibility'. *Canadian Journal of Philosophy* 38 (2): 165–99.

Bok, Sissela. 2004. 'Rethinking the WHO Definition of Health'. Harvard Center for Population and Developmental Studies: Working Paper Series, Vol. 14, No. 7. Available from: http://www.ressma.com/Documentation/BIBLIO/DOCUMENTS%20GENERAUX/WHODEFINITION_HEALTH.pdf.

Boorse, Christopher. 1977. 'Health as a Theoretical Concept'. *Philosophy of Science* 44 (4): 542–73.

Boorse, Christopher. 1997. 'A Rebuttal on Health'. In James M. Humber and Robert F. Almeder, eds, *What Is Disease?*, 1–134. Totowa, NJ: Humana Press.

Borges, Jorge Luis. 1998. 'On Exactitude in Science'. In *Collected Fictions*, translated by Andrew Hurley, 325. New York: Penguin.

Bostrom, Nick. 2005. 'The Fable of the Dragon Tyrant'. *Journal of Medical Ethics* 31 (5): 273–7.

Bovens, Luke, and Nancy Cartwright. 2010. 'Measuring the Impact of Philosophy'. Science and Technology Parliamentary Committee Papers on Research Funding Cuts. Available from: http://www.publications.parliament.uk/pa/cm200910/cmselect/cmsctech/memo/spendingcuts/uc8302.htm.

Box, George E. P., and Norman R. Draper. 1987. *Empirical Model-Building and Response Surfaces*. New York: John Wiley & Sons.

Boyd, Kenneth M. 2000. 'Disease, Illness, Sickness, Health, Healing and Wholeness: Exploring Some Elusive Concepts'. *Medical Humanities* 26 (1): 9–17.

Braveman, Paula A., Catherine Cubbin, Susan Egerter, Sekai Chideya, Kristen S. Marchi, Marilyn Metzler, and Samuel Posner. 2005. 'Socioeconomic Status in Health Research: One Size Does Not Fit All'. *JAMA* 294 (22): 2879–88.

Brennan, Jason. 2018. 'A Libertarian Case for Mandatory Vaccination'. *Journal of Medical Ethics* 44 (1): 37–43.

Brodersen, John, Lisa M. Schwartz, and Steven Woloshin. 2014. 'Overdiagnosis: How Cancer Screening Can Turn Indolent Pathology into Illness'. *APMIS* 122 (8): 683–9.

Broome, John. 1994. 'Discounting the Future'. *Philosophy and Public Affairs* 23 (2): 128–56.

Bush, Vannevar. 1945. *Science: The Endless Frontier*. Washington, DC: United States Government Printing Office.

Butland, Bryony, Susan Jebb, Peter Kopelman, K. McPherson, S. Thomas, J. Mardell, V. Parry, and others. 2007. *Tackling Obesities: Future Choices. Project Report*. London: Government Office for Science.

Butler, Samuel. 1910. *Erewhon: or, Over the Range*. 3rd edn. New York: E. P. Dutton & Company. Available from: https://www.gutenberg.org/files/1906/1906-h/1906-h.htm.

Callahan, Daniel. 1973. 'The WHO Definition of "health"'. *The Hastings Center Studies* 1 (3): 77–87.

Callahan, Daniel. 2013. 'Obesity: Chasing an Elusive Epidemic'. *Hastings Center Report* 43 (1): 34–40.

Campbell, Marion K., Gilda Piaggio, Diana R. Elbourne, and Douglas G. Altman. 2012. 'Consort 2010 Statement: Extension to Cluster Randomised Trials'. *BMJ* 345: e5661.

Campos-Matos, Ines, Seema Mandal, Julie Yates, Mary R. Ramsay, James Wilson, and Wei Shen Lim. 2021. 'Maximising Benefit, Reducing Inequalities and Ensuring Deliverability: Prioritisation of COVID-19 Vaccination in the UK'. *The Lancet Regional Health Europe* 2. Available from: https://www.thelancet.com/journals/lanepe/article/PIIS2666-7762(20)30021-1/fulltext.

Caplan, Arthur L. 1983. 'Can Applied Ethics Be Effective in Health Care and Should It Strive to Be?' *Ethics* 93 (2): 311–19.

Cappelen, Alexander W., and Ole F. Norheim. 2005. 'Responsibility in Health Care: A Liberal Egalitarian Approach'. *Journal of Medical Ethics* 31 (8): 476–80.

Cartwright, Nancy. 2007. 'Are RCTs the Gold Standard?' *BioSocieties* 2 (1): 11–20.

Cartwright, Nancy. 2013. 'Knowing What We Are Talking About: Why Evidence Doesn't Always Travel'. *Evidence & Policy: A Journal of Research, Debate and Practice* 9 (1): 97–112.

Cartwright, Nancy, and Jeremy Hardie. 2012. *Evidence-Based Policy: A Practical Guide to Doing It Better*. New York: Oxford University Press.

Cassini, Alessandro, Liselotte Diaz Högberg, Diamantis Plachouras, Annalisa Quattrocchi, Ana Hoxha, Gunnar Skov Simonsen, Mélanie Colomb-Cotinat, et al. 2019. 'Attributable Deaths and Disability-Adjusted Life-Years Caused by Infections with Antibiotic-Resistant Bacteria in the EU and the European Economic Area in 2015: A Population-Level Modelling Analysis'. *The Lancet Infectious Diseases* 19 (1): 56–66.

Chalmers, Iain. 2003. 'Trying to Do More Good than Harm in Policy and Practice: The Role of Rigorous, Transparent, Up-to-Date Evaluations'. *The ANNALS of the American Academy of Political and Social Science* 589 (1): 22–40.

Chan, Margaret. 2008. 'Finishing the Job of Polio Eradication. Speech to Rotary International Convention: The Rotary Advantage in Polio Eradication'. Available from: https://www.who.int/director-general/speeches/detail/finishing-the-job-of-polio-eradication.

Chan, Margaret. 2016. 'WHO Director-General Briefs UN on Antimicrobial Resistance'. World Health Organization. Available from: https://www.who.int/director-general/speeches/detail/who-director-general-briefs-un-on-antimicrobial-resistance.

Chantler, Tracey, Emilie Karafillakis, and James Wilson. 2019. 'Vaccination: Is There a Place for Penalties for Non-Compliance?' *Applied Health Economics and Health Policy* 17 (3): 265–71.

Charlton, Victoria, Peter Littlejohns, Katharina Kieslich, Polly Mitchell, Benedict Rumbold, Albert Weale, James Wilson, and Annette Rid. 2017. 'Cost Effective but Unaffordable: An Emerging Challenge for Health Systems'. *BMJ* 356: j1402.

Claeys, Gregory. 2013. *Mill and Paternalism*. Cambridge: Cambridge University Press.

Cochi, S., and W. R. Dowdle. 2011. 'The Eradication of Infectious Diseases. Understanding the Lessons and Advancing Experience'. In S. Cochi and W. R. Dowdle, eds, *Disease Eradication in the 21st Century: Implications for Global Health*, 1–10. Cambridge, MA: MIT Press.

Cochrane, Archibald Leman. 1972. *Effectiveness and Efficiency: Random Reflections on Health Services*. London: Nuffield Provincial Hospitals Trust London.

Coggon, John. 2018. 'The Nanny State Debate: A Place Where Words Don't Do Justice'. Faculty of Public Health. Available from: https://www.fph.org.uk/media/2009/fph-nannystatedebate-report-final.pdf.

Cohen, Gerald A. 1989. 'On the Currency of Egalitarian Justice'. *Ethics* 99 (4): 906–44.

Cohen, Gerald A. 1995. *Self-Ownership, Freedom, and Equality*. Cambridge: Cambridge University Press.

Cohen, Gerald A. 2009. *Rescuing Justice and Equality*. Cambridge, MA: Harvard University Press.

Cohen Christofidis, Miriam. 2004. 'Talent, Slavery and Envy in Dworkin's Equality of Resources'. *Utilitas* 16 (3): 267–87.

Commission on the Social Determinants of Health. 2008. *Closing the Gap in a Generation: Health Equity Through Action on the Social Determinants of Health. Final Report of the Commission on Social Determinants of Health*. Geneva: World Health Organization.

Cookson, Richard, Christopher McCabe, and Aki Tsuchiya. 2008. 'Public Healthcare Resource Allocation and the Rule of Rescue'. *Journal of Medical Ethics* 34 (7): 540–4.

Craven, Luke K. 2017. 'System Effects: A Hybrid Methodology for Exploring the Determinants of Food In/Security'. *Annals of the American Association of Geographers* 107 (5): 1011–27.

Cribb, Alan. 2010. 'Translational Ethics? The Theory-Practice Gap in Medical Ethics'. *Journal of Medical Ethics* 36 (4): 207–10.

Cribb, Alan. 2011. 'Beyond the Classroom Wall: Theorist-Practitioner Relationships and Extra-Mural Ethics'. *Ethical Theory and Moral Practice* 14 (4): 383–96.

Dancy, Jonathan. 1985. 'The Role of Imaginary Cases in Ethics'. *Pacific Philosophical Quarterly* 66 (1–2): 141–53.

Daniels, Norman. 1994. 'Four Unsolved Rationing Problems: A Challenge'. *The Hastings Center Report* 24 (4): 27–9.

Daniels, Norman. 2006. 'Equity and Population Health: Toward a Broader Bioethics Agenda'. *The Hastings Center Report* 36 (4): 22–35.

Daniels, Norman. 2007. *Just Health: Meeting Health Needs Fairly*. New York: Cambridge University Press.

Daniels, Norman. 2015. 'Can There Be Moral Force in Favoring an Identified over a Statistical Life?' In I. Glenn Cohen, Norman Daniels, and Nir Eyal, eds, *Identified Versus Statistical Lives: An Interdisciplinary Perspective*, 110–23. New York: Oxford University Press.

Daniels, Norman. 2018. 'Reflective Equilibrium'. In *The Stanford Encyclopedia of Philosophy*, edited by Edward N. Zalta, Fall 2018. Metaphysics Research Lab, Stanford University. Available from: https://plato.stanford.edu/archives/fall2018/entries/reflective-equilibrium/.

Darwall, Stephen. 2006. 'The Value of Autonomy and Autonomy of the Will'. *Ethics* 116 (2): 263–84.

Dave, Chintan V., Aaron S. Kesselheim, Erin R. Fox, Peihua Qiu, and Abraham Hartzema. 2017. 'High Generic Drug Prices and Market Competition'. *Annals of Internal Medicine* 167 (3): 145–51.

Davis, Pamela B. 2006. 'Cystic Fibrosis Since 1938'. *American Journal of Respiratory and Critical Care Medicine* 173 (5): 475–82.

Dawson, Angus. 2011. 'Resetting the Parameters: Public Health as the Foundation for Public Health Ethics'. In Angus Dawson, ed., *Public Health Ethics: Key Concepts and Issues in Policy and Practice*, 1–19. Cambridge: Cambridge University Press.

Dawson, Angus, and Marcel Verweij. 2008. 'The Steward of the Millian State'. *Public Health Ethics* 1 (3): 193–5.

D'Costa, Vanessa M., Christine E. King, Lindsay Kalan, Mariya Morar, Wilson W. L. Sung, Carsten Schwarz, Duane Froese, et al. 2011. 'Antibiotic Resistance Is Ancient'. *Nature* 477 (7365): 457–61.

de Marneffe, Peter. 2006. 'Avoiding Paternalism'. *Philosophy and Public Affairs* 34 (1): 68–94.

Deaton, Angus. 2003. 'Health, Inequality, and Economic Development'. *Journal of Economic Literature* 41 (1): 113–58.

Deaton, Angus, and Nancy Cartwright. 2018. 'Understanding and Misunderstanding Randomized Controlled Trials'. *Social Science & Medicine* 210: 2–21.

Dees, Richard H. 2017. 'Public Health and Normative Public Goods'. *Public Health Ethics* 11 (1): 20–6.

Delamater, Paul L., Erica J. Street, Timothy F. Leslie, Y. Tony Yang, and Kathryn H. Jacobsen. 2019. 'Complexity of the Basic Reproduction Number (R0)'. *Emerging Infectious Diseases* 25 (1): 1–4.

Dennett, Daniel C. 2006. 'Higher-Order Truths about Chmess'. *Topoi* 25 (1–2): 39–41.

Dewey, John. 1917. 'The Need for a Recovery of Philosophy'. In *Creative Intelligence: Essays in the Pragmatic Attitude*. New York: Henry Holt and Co. Available from: http://www.gutenberg.org/ebooks/33727.

Dewey, John, and James Tufts. 1981. 'Ethics'. In *The Later Works, 1925–1953*, volume 7, edited by J. A. Boydston. Carbondale: Southern Illinois University Press.

Doll, Richard, and A. Bradford Hill. 2004. 'The Mortality of Doctors in Relation to Their Smoking Habits: A Preliminary Report. 1954'. *BMJ (Clinical Research Edn)* 328 (7455): 1529–33.

Drlica, Karl, and David S. Perlin. 2011. *Antibiotic Resistance: Understanding and Responding to an Emerging Crisis*. Upper Saddle River, NJ: FT Press.

Drolet, Brian C., and Nancy M. Lorenzi. 2011. 'Translational Research: Understanding the Continuum from Bench to Bedside'. *Translational Research* 157 (1): 1–5.

Dumit, Joseph. 2006. 'Illnesses You Have to Fight to Get: Facts as Forces in Uncertain, Emergent Illnesses'. *Social Science & Medicine* 62 (3): 577–90.

Dunphy, Kilian. 2014. 'Herpes Genitalis and the Philosopher's Stance'. *Journal of Medical Ethics* 40 (12): 793–7.

Dworkin, Gerald. 1972. 'Paternalism'. *The Monist* 56 (1): 64–84.

Dworkin, Gerald. 2020. 'Paternalism'. In *The Stanford Encyclopedia of Philosophy*, edited by Edward N. Zalta, Summer 2020. Metaphysics Research Lab, Stanford University. Available from: https://plato.stanford.edu/archives/sum2020/entries/paternalism/.

Dworkin, Ronald. 1977. *Taking Rights Seriously*. London: Duckworth.

Dworkin, Ronald. 1981a. 'What Is Equality? Part 1: Equality of Welfare'. *Philosophy and Public Affairs* 10 (3): 185–246.

Dworkin, Ronald. 1981b. 'What Is Equality? Part 2: Equality of Resources'. *Philosophy and Public Affairs* 10 (4): 283–345.

Dyer, Owen. 2020. 'Covid-19: Many Poor Countries Will See Almost No Vaccine Next Year, Aid Groups Warn'. *BMJ* 371 (December): m4809.

Easwaran, Eknath. 1986. *The Dhammapada*. London: Routledge & Kegan Paul.

Eddy, David M. 2005. 'Evidence-Based Medicine: A Unified Approach'. *Health Affairs* 24 (1): 9–17.

Edwards, Sarah J. L., and James Wilson. 2012. 'Hard Paternalism, Fairness and Clinical Research: Why Not?' *Bioethics* 26 (2): 68–75.

Elgin, Catherine Z. 2014. 'Fiction as Thought Experiment'. *Perspectives on Science* 22 (2): 221–41.

Elster, Jakob. 2011. 'How Outlandish Can Imaginary Cases Be?' *Journal of Applied Philosophy* 28 (3): 241–58.

Emanuel, Ezekiel J. 2004. 'Ending Concerns about Undue Inducement'. *The Journal of Law, Medicine & Ethics* 32 (1): 100–5.

Emerson, Claudia. 2011. 'The Moral Case for Eradication'. In S. L. Cochi and W. R. Dowdle, eds, *Disease Eradication in the 21st Century: Implications for Global Health*, 103–14. Cambridge, MA: MIT Press.

Faust, Halley S., and Paul T. Menzel. 2011. 'Introduction'. In Halley S. Faust and Paul T. Menzel, eds, *Prevention Vs. Treatment What's the Right Balance?*, 1–31. New York: Oxford University Press.

Feinberg, Joel. 1970. 'The Nature and Value of Rights'. *The Journal of Value Inquiry* 4 (4): 243–60.

Feinberg, Joel. 1984. *The Moral Limits of the Criminal Law. Volume 1: Harm to Others*. New York: Oxford University Press.

Feinberg, Joel. 1986. *The Moral Limits of the Criminal Law. Volume 3: Harm to Self*. New York: Oxford University Press.

Fejerskov, Adam Moe. 2018. 'Development as Resistance and Translation: Remaking Norms and Ideas of the Gates Foundation'. *Progress in Development Studies* 18 (2): 126–43.

Fenner, F., A. J. Hall, and W. R. Dowdle. 1998. 'What Is Eradication?' In W. R. Dowdle and D. R. Hopkins, eds, *The Eradication of Infectious Diseases*, 3–17. New York: John Wiley and Sons.

Ferrario, Alessandra, Guillaume Dedet, Tifenn Humbert, Sabine Vogler, Fatima Suleman, and Hanne Bak Pedersen. 2020. 'Strategies to Achieve Fairer Prices for Generic and Biosimilar Medicines'. *BMJ* 368 (January): l5444.

Fine, Paul, Ken Eames, and David L. Heymann. 2011. '"Herd Immunity": A Rough Guide'. *Clinical Infectious Diseases* 52 (7): 911–16.

Fisher, Ronald A. 1935. *The Design of Experiments*. Edinburgh: Oliver & Boyd.

Flanigan, Jessica. 2013. 'Public Bioethics'. *Public Health Ethics* 6 (2): 170–84.

Flanigan, Jessica. 2014. 'A Defense of Compulsory Vaccination'. *HEC Forum* 26 (1): 5–25.

Flanigan, Jessica. 2017. 'Seat Belt Mandates and Paternalism'. *Journal of Moral Philosophy* 14 (3): 291–314.

Fleming, Alexander. 1980. 'On the Antibacterial Action of Cultures of a Penicillium, with Special Reference to Their Use in the Isolation of b. Influenzae'. *Reviews of Infectious Diseases* 2 (1): 129–39.

Fleming-Dutra, Katherine E., Adam L. Hersh, Daniel J. Shapiro, Monina Bartoces, Eva A. Enns, Thomas M. File, Jonathan A. Finkelstein, et al. 2016. 'Prevalence of Inappropriate Antibiotic Prescriptions Among US Ambulatory Care Visits, 2010-2011'. *JAMA* 315 (17): 1864–73.

Fleurbaey, Marc. 1995. 'Equal Opportunity or Equal Social Outcome?' *Economics & Philosophy* 11 (1): 25–55.

Fleurbaey, Marc, and Alex Voorhoeve. 2013. 'Decide as You Would with Full Information! An Argument Against Ex Ante Pareto'. In Ole F. Norheim, Samia Hurst, Nir Eyal, and Dan Wikler, eds, *Inequalities in Health: Concepts, Measures, and Ethics*, 113–28. New York: Oxford University Press.

Foot, Philippa. 1967. 'The Problem of Abortion and the Doctrine of Double Effect'. *Oxford Review* 5: 5–15.

Fox, John P., Lila Elveback, William Scott, Lael Gatewood, and Eugene Ackerman. 1971. 'Herd Immunity: Basic Concept and Relevance to Public Health Immunization Practices'. *American Journal of Epidemiology* 94 (3): 179–89.

Frey, Bruno S, and Reto Jegen. 2001. 'Motivation Crowding Theory'. *Journal of Economic Surveys* 15 (5): 589–611.

Frick, Johann. 2015. 'Contractualism and Social Risk'. *Philosophy and Public Affairs* 43 (3): 175–223.

Fried, Barbara. 2012a. 'Can Contractualism Save Us from Aggregation?' *Journal of Ethics* 16 (1): 39–66.

Fried, Barbara. 2012b. 'What Does Matter? The Case for Killing the Trolley Problem (or Letting It Die)'. *Philosophical Quarterly* 62 (248): 505–29.

Fried, Barbara. 2019. 'Facing up to Risk'. *Journal of Legal Analysis* 10: 175–98.

Friesen, Phoebe. 2018. 'Personal Responsibility Within Health Policy: Unethical and Ineffective'. *Journal of Medical Ethics* 44 (1): 53–8.

Gates, Melinda French. 2019. *The Moment of Lift: How Empowering Women Changes the World*. New York: Flatiron Books.

Gergel, Tania Louise. 2014. 'Too Similar, Too Different: The Paradoxical Dualism of Psychiatric Stigma'. *The Psychiatric Bulletin* 38 (4): 148–51.

Gericke, C. A., R. Busse, and A. Riesberg. 2005. 'Ethical Issues in Funding Orphan Drug Research and Development'. *Journal of Medical Ethics* 31 (3): 164–8.

Gert, Bernard, and Charles M. Culver. 1976. 'Paternalistic Behavior'. *Philosophy and Public Affairs* 6 (1): 45–57.

Global Commission on HIV and the Law. 2012. *Risks, Rights & Health*. New York: UNDP. Available from: https://hivlawcommission.org/report/.

Glover, Jonathan. 1975. 'It Makes No Difference Whether or Not I Do It'. *Proceedings of the Aristotelian Society* 49: 171–209.

Godin, Benoît. 2006. 'The Linear Model of Innovation: The Historical Construction of an Analytical Framework'. *Science, Technology & Human Values* 31 (6): 639–67.

Goffman, Erving. 1963. *Stigma: Notes on the Management of Spoiled Identity*. Englewood Cliffs, NJ: Prentice-Hall.

Gostin, Lawrence O. 2014. *Global Health Law*. Cambridge, MA: Harvard University Press.

Greenhalgh, Trisha. 2006. *How to Read a Paper: The Basics of Evidence-Based Medicine*. 3rd edn. Oxford: WileyBlackwell.

Greenhalgh, Trisha, and Jill Russell. 2006. 'Reframing Evidence Synthesis as Rhetorical Action in the Policy Making Drama'. *Healthcare Policy* 1 (2): 34–42.

Grill, Kalle. 2009. 'Liberalism, Altruism and Group Consent'. *Public Health Ethics* 2 (2): 146–57.

Groce, Nora Ellen. 1985. *Everyone Here Spoke Sign Language*. Cambridge, MA: Harvard University Press.

Gustafsson, Johan E. 2015. 'Sequential Dominance and the Anti-Aggregation Principle'. *Philosophical Studies* 172 (6): 1593–601.

Guyatt, Gordon H., Andrew D. Oxman, Gunn E. Vist, Regina Kunz, Yngve Falck-Ytter, Pablo Alonso-Coello, and Holger J. Schünemann. 2008. 'GRADE: An Emerging Consensus on Rating Quality of Evidence and Strength of Recommendations'. *BMJ (Clinical Research Edn)* 336 (7650): 924–6.

Habermas, Jürgen. 2003. *The Future of Human Nature*. Cambridge: Polity Press.

Hacking, Ian. 2007. 'Kinds of People: Moving Targets'. *Proceedings of the British Academy* 151: 285–318.

Hájek, Alan. 2007. 'The Reference Class Problem Is Your Problem Too'. *Synthese* 156 (3): 563–85.

Halstead, John. 2016. 'The Numbers Always Count'. *Ethics* 126 (3): 789–802.

Hammersley, Martyn. 2005. 'Is the Evidence-Based Practice Movement Doing More Good Than Harm? Reflections on Iain Chalmers' Case for Research-Based Policy Making and Practice'. *Evidence & Policy: A Journal of Research, Debate and Practice* 1 (January): 85–100.

Hansson, Sven Ove. 2005. 'Extended Antipaternalism'. *Journal of Medical Ethics* 31 (2): 97–100.

Harker, Rachel. 2018. *NHS Funding and Expenditure: Briefing Paper CBP0724*. London: House of Commons Library. Available from: http://researchbriefings.files.parliament.uk/documents/SN00724/SN00724.pdf.

Hart, H. L. A. 1982. *Essays on Bentham: Studies in Jurisprudence and Political Theory*. Oxford: Clarendon Press.

Hatzenbuehler, Mark L., Jo C. Phelan, and Bruce G. Link. 2013. 'Stigma as a Fundamental Cause of Population Health Inequalities'. *American Journal of Public Health* 103 (5): 813–21.

Hausman, Daniel. 2007. 'What's Wrong with Health Inequalities?' *Journal of Political Philosophy* 15 (1): 46–66.

Hausman, Daniel. 2015. *Valuing Health: Well-Being, Freedom, and Suffering*. New York: Oxford University Press.

Hausman, Daniel, Yukiko Asada, and Thomas Hedemann. 2002. 'Health Inequalities and Why They Matter'. *Health Care Analysis* 10: 177–91.

Hawton, Keith, Sue Simkin, Jonathan Deeks, Jayne Cooper, Amy Johnston, Keith Waters, Morag Arundel, et al. 2004. 'UK Legislation on Analgesic Packs: Before and After Study of Long Term Effect on Poisonings'. *BMJ* 329 (7474): 1076.

Hayek, Friedrich A. 1944. *The Road to Serfdom*. London: Routledge.

Hayek, Friedrich A. 1967. 'The Theory of Complex Phenomena'. In *Studies in Philosophy, Politics and Economics*, 22–42. Chicago, IL: University of Chicago Press.

Haynes, Laura, Owain Service, Ben Goldacre, and David Torgerson. 2012. 'Test, Learn, Adapt: Developing Public Policy with Randomised Controlled Trials'. London: Cabinet Office Behavioural Insights Team. Available from: https://assets.publishing.service.gov. uk/government/uploads/system/uploads/attachment_data/file/62529/TLA-1906126.pdf.

Head, Megan L., Luke Holman, Rob Lanfear, Andrew T. Kahn, and Michael D. Jennions. 2015. 'The Extent and Consequences of p-Hacking in Science'. *PLoS Biology* 13 (3): e1002106.

Hemming, K., S. Eldridge, G. Forbes, C. Weijer, and M. Taljaard. 2017. 'How to Design Efficient Cluster Randomised Trials'. *BMJ* 358 (July): j3064.

Hermansson, Helene, and Sven Ove Hansson. 2007. 'A Three-Party Model Tool for Ethical Risk Analysis'. *Risk Management* 9 (3): 129–44.

Hill, Austin Bradford. 1965. 'The Environment and Disease: Association or Causation?' *Proceedings of the Royal Society of Medicine* 58 (5): 295–300.

Hofmann, Bjørn. 2016. 'Disease, Illness, and Sickness'. In Miriam Solomon, Jeremy R. Simon, and Harold Kincaid, eds, *The Routledge Companion to Philosophy of Medicine*, 16–26. Abingdon: Routledge.

Hofmann, Bjørn. 2018. 'Getting Personal on Overdiagnosis: On Defining Overdiagnosis from the Perspective of the Individual Person'. *Journal of Evaluation in Clinical Practice* 24 (5): 983–7.

Hollis, Adrian. 2006. 'Drugs for Rare Diseases: Paying for Innovation'. In Charles M. Beach, Richard P. Chaykowski, and S. E. D. Shortt, eds, *Health Services Restructuring in Canada: New Evidence and New Directions*, 155–78. Montreal: Queen's University Press.

Holm, Soren. 1995. 'Not Just Autonomy—the Principles of American Biomedical Ethics'. *Journal of Medical Ethics* 21 (6): 332–8.

Horne, L. Chad. 2019. 'Public Health, Public Goods, and Market Failure'. *Public Health Ethics* 12 (3): 287–92.

Husak, Douglas. 2003. 'Legal Paternalism'. In Hugh LaFollette, ed., *The Oxford Handbook of Practical Ethics*, 387–8. Oxford: Oxford University Press.

Hutchinson, Phil, and Rageshri Dhairyawan. 2017. 'Shame, Stigma, HIV: Philosophical Reflections'. *Medical Humanities* 43 (4): 225–30.

Institute of Medicine and National Research Council. 2013. *U.S. Health in International Perspective: Shorter Lives, Poorer Health*. Edited by Steven H. Woolf and Laudan Aron. Washington, DC: The National Academies Press.

Ioannidis, John P. A. 2005. 'Why Most Published Research Findings Are False'. *PLoS Medicine* 2 (8): e124.

Jackson, Charlotte, Emilia Vynnycky, Jeremy Hawker, Babatunde Olowokure, and Punam Mangtani. 2013. 'School Closures and Influenza: Systematic Review of Epidemiological Studies'. *BMJ Open* 3 (2): e002149.

Jacobs, Jane. 1961. *The Death and Life of Great American Cities*. New York: Vintage.

James, William. 1897. 'The Will to Believe'. In *The Will to Believe and Other Essays in Popular Philosophy*. New York: Longmans, Green and Co. Available from: https://www. gutenberg.org/files/26659/26659-h/26659-h.htm.

James, William. 1907. *Pragmatism: A New Name for Some Old Ways of Thinking*. New York: Longmans, Green and Co. Available from: https://www.gutenberg.org/ebooks/5116.

Jansen, Kathrin U., Charles Knirsch, and Annaliesa S. Anderson. 2018. 'The Role of Vaccines in Preventing Bacterial Antimicrobial Resistance'. *Nature Medicine* 24 (1): 10–19.

Jefferson, Thomas. 1806. 'Letter to Edward Jenner'. Available from: https://www.loc.gov/resource/mtj1.036_0006_0006/.

Jellinger, Kurt A., and Amos D. Korczyn. 2018. 'Are Dementia with Lewy Bodies and Parkinson's Disease Dementia the Same Disease?' *BMC Medicine* 16 (1): 1–16.

Jenni, Karen, and George Loewenstein. 1997. 'Explaining the Identifiable Victim Effect'. *Journal of Risk and Uncertainty* 14 (3): 235–57.

Jennings, Bruce. 2009. 'Public Health and Liberty: Beyond the Millian Paradigm'. *Public Health Ethics* 2 (2): 123–34.

John, Stephen D. 2013. 'Cancer Screening, Risk Stratification and the Ethics of Apt Categorisation: A Case Study'. In Daniel Strech, Irene Hirschberg, and Georg Marckmann, eds, *Ethics in Public Health and Health Policy: Concepts, Methods, Case Studies*, 141–52. Dordrecht: Springer Netherlands.

John, Stephen D. 2014. 'Risk, Contractualism, and Rose's "Prevention Paradox"'. *Social Theory and Practice* 40 (1): 28–50.

John, Stephen D. 2020. 'The Ethics of Lockdown: Communication, Consequences, and the Separateness of Persons'. *Kennedy Institute of Ethics Journal*. Available from: https://kiej.georgetown.edu/ethics-of-lockdown-special-issue/.

Jonas, Monique. 2016. 'Child Health Advice and Parental Obligation: The Case of Safe Sleep Recommendations and Sudden Unexpected Death in Infancy'. *Bioethics* 30 (2): 129–38.

Jonas, Monique, and Riripeti Haretuku. 2016. 'Reducing Sudden Infant Death Syndrome in a Culturally Diverse Society: The New Zealand Cot Death Study and National Cot Death Prevention Programme'. In Drue H. Barrett, Leonard W. Ortmann, Angus Dawson, Carla Saenz, Andreas Reis, and Gail Bolan, eds, *Public Health Ethics: Cases Spanning the Globe*, 211–16. Cham: Springer Open.

Jones, David S., and Scott H. Podolsky. 2015. 'The History and Fate of the Gold Standard'. *The Lancet* 385 (9977): 1502–3.

Jonsen, Albert R. 1986. 'Bentham in a Box: Technology Assessment and Health Care Allocation'. *Law, Medicine and Health Care* 14: 172–4.

Judson, Olivia P. 2017. 'The Energy Expansions of Evolution'. *Nature Ecology & Evolution* 1 (138): 1–9.

Judt, Tony. 2009. 'What Is Living and What Is Dead in Social Democracy?' *New York Review of Books*, December. Available from: https://www.nybooks.com/articles/2009/12/17/what-is-living-and-what-is-dead-in-social-democrac.

Judt, Tony. 2011. *Ill Fares the Land: A Treatise on Our Present Discontents*. London: Penguin.

Kagan, Shelly. 1988. 'The Additive Fallacy'. *Ethics* 99 (1): 5–31.

Kamm, F. M. 1993. *Morality, Mortality, Volume I: Death and Whom to Save from It*. Oxford: Oxford University Press.

Kamm, F. M. 1996. *Morality, Mortality, Volume II: Rights, Duties, and Status*. New York: Oxford University Press.

Kamm, F. M. 2006. *Intricate Ethics: Rights, Responsibilities, and Permissible Harm*. New York: Oxford University Press.

Katz, Jay. 2002. *The Silent World of Doctor and Patient*. Baltimore, MD: Johns Hopkins University Press.

Kawachi, Ichiro, S. V. Subramanian, and Naomar Almeida-Filho. 2002. 'A Glossary for Health Inequalities'. *Journal of Epidemiology & Community Health* 56 (9): 647–52.

Keen, John D. 2010. 'Promoting Screening Mammography: Insight or Uptake?' *Journal of the American Board of Family Medicine* 23 (6): 775–82.

Keesing, Felicia, Lisa K. Belden, Peter Daszak, Andrew Dobson, C. Drew Harvell, Robert D. Holt, Peter Hudson, et al. 2010. 'Impacts of Biodiversity on the Emergence and Transmission of Infectious Diseases'. *Nature* 468 (7324): 647–52.

Kelleher, J. Paul. 2013. 'Prevention, Rescue and Tiny Risks'. *Public Health Ethics* 6 (3): 252–61.

Kelly, Michael P., and Federica Russo. 2018. 'Causal Narratives in Public Health: The Difference Between Mechanisms of Aetiology and Mechanisms of Prevention in Non-Communicable Diseases'. *Sociology of Health & Illness* 40 (1): 82–99.

Kennedy, David A., and Andrew F. Read. 2017. 'Why Does Drug Resistance Readily Evolve but Vaccine Resistance Does Not?' *Proceedings of the Royal Society B: Biological Sciences* 284 (March): 20162562.

Kennedy, David A., and Andrew F. Read. 2018. 'Why the Evolution of Vaccine Resistance Is Less of a Concern Than the Evolution of Drug Resistance'. *Proceedings of the National Academy of Sciences* 115 (51): 12878–86.

Keogh, Ruth H., Rhonda Szczesniak, David Taylor-Robinson, and Diana Bilton. 2018. 'Up-to-Date and Projected Estimates of Survival for People with Cystic Fibrosis Using Baseline Characteristics: A Longitudinal Study Using UK Patient Registry Data'. *Journal of Cystic Fibrosis* 17 (2): 218–27.

Kingma, Elselijn. 2007. 'What Is It to Be Healthy?' *Analysis* 67 (2): 128–33.

Kitcher, Philip. 2011a. *Science in a Democratic Society*. New York: Prometheus Books.

Kitcher, Philip. 2011b. *The Ethical Project*. Cambridge, MA: Harvard University Press.

Kitcher, Philip. 2011c. 'Philosophy Inside Out'. *Metaphilosophy* 42 (3): 248–60.

Kitcher, Philip. 2012. *Preludes to Pragmatism: Toward a Reconstruction of Philosophy*. New York: Oxford University Press.

Klepac, Peter, C. Jessica E. Metcalf, Angela R. McLean, and Katie Hampson. 2013. 'Introduction: Towards the Endgame and Beyond: Complexities and Challenges for the Elimination of Infectious Diseases'. *Philosophical Transactions: Biological Sciences* 368 (1623): 1–12.

Kline, Stephen, and Nathan Rosenberg. 1986. 'An Overview of Innovation'. In Ralph Landau and Nathan Rosenberg, eds, *The Positive Sum Strategy: Harnessing Technology for Economic Growth*, 275–306. Washington, DC: National Academies Press.

Koestler, Arthur. 1970. 'Beyond Atomism and Holism—the Concept of the Holon'. *Perspectives in Biology and Medicine* 13 (2): 131–54.

Kymlicka, Will. 2002. *Contemporary Political Philosophy: An Introduction*. New York: Oxford University Press.

Ladyman, James, James Lambert, and Karoline Wiesner. 2013. 'What Is a Complex System?' *European Journal for Philosophy of Science* 3 (1): 33–67.

Laland, Kevin, Blake Matthews, and Marcus W. Feldman. 2016. 'An Introduction to Niche Construction Theory'. *Evolutionary Ecology* 30 (2): 191–202.

Landrigan, Philip J., Richard Fuller, Nereus J. R. Acosta, Olusoji Adeyi, Robert Arnold, Abdoulaye Bibi Baldé, Roberto Bertollini, et al. 2017. 'The Lancet Commission on Pollution and Health'. *The Lancet* 391 (10119): 462–512.

Le Fanu, James. 2000. *The Rise and Fall of Modern Medicine*. London: Abacus.

Letelier, Orlando. 1976. 'The Chicago Boys in Chile: Economic Freedom's Awful Toll'. *The Nation*. Reprinted in *The Nation*, 21 September 2016. Available from: https://www.thenation.com/article/archive/the-chicago-boys-in-chile-economic-freedoms-awful-toll/.

Levin, Kelly, Benjamin Cashore, Steven Bernstein, and Graeme Auld. 2012. 'Overcoming the Tragedy of Super Wicked Problems: Constraining Our Future Selves to Ameliorate Global Climate Change'. *Policy Sciences* 45 (2): 123–52.

Lewis, Paul. 2017. 'The Ostroms and Hayek as Theorists of Complex Adaptive Systems: Commonality and Complementarity'. In Paul Dragos Aligica, Paul Lewis, and Virgil H. Storr, eds, *The Austrian and Bloomington Schools of Political Economy*, 35–66. Bingley: Emerald Group Publishing Limited.

Liebisch, Tara Cubel, Jörn Stenger, and Joachim Ullrich. 2019. 'Understanding the Revised SI: Background, Consequences, and Perspectives'. *Annalen Der Physik* 531 (5): 1800339.

Ligon, B. Lee. 2004. 'Penicillin: Its Discovery and Early Development'. *Seminars in Pediatric Infectious Diseases* 15 (1): 52–7.

Lindblom, Charles. E. 1959. 'The Science of "Muddling Through"'. *Public Administration Review* 19 (2): 79–88.

Link, Bruce G., and Jo C. Phelan. 2001. 'Conceptualizing Stigma'. *Annual Review of Sociology* 27 (1): 363–85.

Lipsitch, Marc, and George R. Siber. 2016. 'How Can Vaccines Contribute to Solving the Antimicrobial Resistance Problem?' *mBio* 7 (3): e00428-16.

Lipsky, Michael. 2010. *Street-Level Bureaucracy, 30th Anniversary Edition: Dilemmas of the Individual in Public Service*. New York: Russell Sage Foundation.

Littmann, Jasper, and A. M. Viens. 2015. 'The Ethical Significance of Antimicrobial Resistance'. *Public Health Ethics* 8 (3): 209–24.

Littmann, Jasper, A. M. Viens, and Diego S. Silva. 2019. 'The Super-Wicked Problem of Antimicrobial Resistance'. In Euzebiusz Jamrozik and Michael Selgelid, eds, *Ethics and Drug Resistance: Collective Responsibility for Global Public Health*, 421–43. Cham: Springer International Publishing.

Liu, Jenny, Sepideh Modrek, Roly Gosling, and Richard G. A. Feachem. 2013. 'Malaria Eradication: Is It Possible? Is It Worth It? Should We Do It?' *The Lancet Global Health* 1 (1): e2–e3.

Loader, Ian, and Neil Walker. 2007. *Civilizing Security*. Cambridge: Cambridge University Press.

Lorenz, Edward N. 1963. 'Deterministic Nonperiodic Flow'. *Journal of the Atmospheric Sciences* 20 (2): 130–41.

McCabe, Christopher. 2005. 'Orphan Drugs and the NHS: Should We Value Rarity?' *BMJ* 331 (7523): 1016–19.

McGarvey, Darren. 2018. *Poverty Safari: Understanding the Anger of Britain's Underclass*. London: Picador.

McKie, John, and Jeff Richardson. 2003. 'The Rule of Rescue'. *Social Science & Medicine* 56 (12): 2407–19.

Manson, Neil C., and Onora O'Neill. 2007. *Rethinking Informed Consent in Bioethics*. Cambridge: Cambridge University Press.

Marincola, Francesco M. 2003. 'Translational Medicine: A Two-Way Road'. *Journal of Translational Medicine* 1 (1): 1. Available from: https://translational-medicine.biomed-central.com/articles/10.1186/1479-5876-1-1.

Marmot, Michael. 2005. 'Social Determinants of Health Inequalities'. *The Lancet* 365 (9464): 1099–104.

Marmot, Michael, D. G. Altman, D. A. Cameron, J. A. Dewar, S. G. Thompson, and Maggie Wilcox. 2013. 'The Benefits and Harms of Breast Cancer Screening: An Independent Review'. *British Journal of Cancer* 108 (11): 2205–40.

Marmot, Michael, Geoffrey Rose, M. J. Shipley, and P. J. Hamilton. 1978. 'Employment Grade and Coronary Heart Disease in British Civil Servants'. *Journal of Epidemiology and Community Health* 32: 244–9.

Martins, Carlos, Maciek Godycki-Cwirko, Bruno Heleno, and John Brodersen. 2018. 'Quaternary Prevention: Reviewing the Concept'. *European Journal of General Practice* 24 (1): 106–11.

May, Robert M. 1976. 'Simple Mathematical Models with Very Complicated Dynamics'. *Nature* 261: 459–67.

Meadows, Donella H. 2008. *Thinking in Systems: A Primer*. Edited by Diana Wright. London: Earthscan.

Meadows, Donella H., Dennis L. Meadows, William W. III Behrens, and Jørgen Randers. 1972. *The Limits to Growth: A Report for the Club of Rome's Project on the Predicament of Mankind*. New York: Universe Books.

Medical Research Council. 1948. 'Streptomycin Treatment of Pulmonary Tuberculosis: A Medical Research Council Investigation'. *BMJ* 2 (4582): 769–82.

Merton, Robert K. 1948. 'The Self-Fulfilling Prophecy'. *The Antioch Review* 8 (2): 193–210.

Merton, Robert K. 1968. 'The Matthew Effect in Science'. *Science* 159 (3810): 56–63.

Met Office. 2020. 'Numerical Weather Prediction Models'. Available from: https://www.metoffice.gov.uk/research/approach/modelling-systems/unified-model/weather-forecasting.

Mill, John Stuart. 1977 [1860]. *On Liberty*. In *The Collected Works of John Stuart Mill, Volume XVIII—Essays on Politics and Society Part I*, ed. John M. Robson, Introduction by Alexander Brady, 213–310. London: Routledge and Kegan Paul. Available from: https://oll.libertyfund.org/title/robson-the-collected-works-of-john-stuart-mill-volume-xviii-essays-on-politics-and-society-part-i.

Mill, John Stuart. 1985 [1874]. *On Nature*. In *The Collected Works of John Stuart Mill, Volume X—Essays on Ethics, Religion, and Society*, ed. John M. Robson, Introduction by F. E. L. Priestley, 373–402. London: Routledge and Kegan Paul. Available from: https://oll.libertyfund.org/title/mill-the-collected-works-of-john-stuart-mill-volume-x-essays-on-ethics-religion-and-society.

Millar, Michael. 2011. 'Can Antibiotic Use Be Both Just and Sustainable…Or Only More or Less So?' *Journal of Medical Ethics* 37 (3): 153–7.

Millar, Michael. 2012. 'Constraining the Use of Antibiotics: Applying Scanlon's Contractualism'. *Journal of Medical Ethics* 38 (8): 465–9.

Mitchell, Polly. 2018. 'The Construction of Well-Being'. PhD Thesis, University College London. Available from: https://discovery.ucl.ac.uk/id/eprint/10064726/.

Mitchell, Polly, and Anna Alexandrova. 2020. 'Well-Being and Pluralism'. *Journal of Happiness Studies*. DOI: 10.1007/s10902-020-00323-8.

Morris, John C. 2000. 'The Nosology of Dementia'. *Neurologic Clinics* 18 (4): 773–88.

Mounk, Yascha. 2017. *The Age of Responsibility: Luck, Choice, and the Welfare State*. Cambridge: MA: Harvard University Press.

Mueller, Charles W., and Toby L. Parcel. 1981. 'Measures of Socioeconomic Status: Alternatives and Recommendations'. *Child Development* 52 (1): 13–30.

Munro, Eileen. 2005. 'A Systems Approach to Investigating Child Abuse Deaths'. *British Journal of Social Work* 35 (4): 531–46.

Murray, Christopher J. 1994. 'Quantifying the Burden of Disease: The Technical Basis for Disability-Adjusted Life Years'. *Bulletin of the World Health Organization* 72 (3): 429–45.

Murray, Christopher J., Emmanuela E Gakidou, and Julio Frenk. 1999. 'Health Inequalities and Social Group Differences: What Should We Measure?' *Bulletin of the World Health Organization* 77 (7): 537–43.

Nagel, Thomas. 1979. 'Equality'. In *Mortal Questions*, 106–27. Cambridge: Cambridge University Press.

Navin, Mark Christopher, and Mark Aaron Largent. 2017. 'Improving Nonmedical Vaccine Exemption Policies: Three Case Studies'. *Public Health Ethics* 10 (3): 225–34.

Nissenbaum, Helen. 2009. *Privacy in Context: Technology, Policy, and the Integrity of Social Life*. Stanford, CA: Stanford University Press.

Nozick, Robert. 1974. *Anarchy, State and Utopia*. New York: Basic Books.

Nussbaum, Martha C. 2000. *Women and Human Development: The Capabilities Approach*. Cambridge: Cambridge University Press.

Nussbaum, Martha C. 2011. *Creating Capabilities*. Cambridge, MA: Harvard University Press.

Nys, Thomas R. V. 2008. 'Paternalism in Public Health Care'. *Public Health Ethics* 1 (1): 64–72.

Office for National Statistics. 2016. 'Smoking Inequalities in England, 2016'. Available from: https://www.ons.gov.uk/peoplepopulationandcommunity/healthandsocialcare/drugusealcoholandsmoking/adhocs/008181smokinginequalitiesinengland2016.

Office for National Statistics. 2019. 'Healthcare Expenditure, UK Health Accounts—Office for National Statistics'. Available from: https://www.ons.gov.uk/peoplepopulationand-community/healthandsocialcare/healthcaresystem/bulletins/ukhealthaccounts/2017.

Omer, Saad B., Walter A. Orenstein, and Jeffrey P. Koplan. 2013. 'Go Big and Go Fast—Vaccine Refusal and Disease Eradication'. *New England Journal of Medicine* 368 (15): 1374–6.

O'Neill, Martin. 2008. 'What Should Egalitarians Believe?' *Philosophy and Public Affairs* 36 (2): 119–56.

O'Neill, Onora. 2002a. *Autonomy and Trust in Bioethics*. Cambridge: Cambridge University Press.

O'Neill, Onora. 2002b. *A Question of Trust: The BBC Reith Lectures 2002*. Cambridge: Cambridge University Press.

Ostfeld, Richard. 2012. *Lyme Disease: The Ecology of a Complex System*. New York: Oxford University Press.

Ostrom, Elinor. 1990. *Governing the Commons: The Evolution of Institutions for Collective Action*. Cambridge: Cambridge University Press.

Ostrom, Elinor. 2010. 'Beyond Markets and States: Polycentric Governance of Complex Economic Systems'. *American Economic Review* 100 (3): 641–72.

Owen, Lesley, Antony Morgan, Alastair Fischer, Simon Ellis, Andrew Hoy, and Michael P. Kelly. 2012. 'The Cost-Effectiveness of Public Health Interventions'. *Journal of Public Health* 34 (1): 37–45.

Page, Scott E. 2018. *The Model Thinker: What You Need to Know to Make Data Work for You*. New York: Basic Books.

Parfit, Derek. 1984. *Reasons and Persons*. Oxford: Oxford University Press.

Parker, Richard, and Peter Aggleton. 2003. 'HIV and AIDS-Related Stigma and Discrimination: A Conceptual Framework and Implications for Action'. *Social Science & Medicine* 57 (1): 13–24.

Parkhurst, Justin O., and Sudeepa Abeysinghe. 2016. 'What Constitutes Good Evidence for Public Health and Social Policy-Making? From Hierarchies to Appropriateness'. *Social Epistemology* 30 (5–6): 665–79.

Parkkinen, Veli-Pekka, Christian Wallmann, Michael Wilde, Brendan Clarke, Phyllis Illari, Michael P. Kelly, Charles Norell, Federica Russo, Beth Shaw, and Jon Williamson. 2018. *Evaluating Evidence of Mechanisms in Medicine: Principles and Procedures*. Cham: Springer.

Pashler, Harold, and Eric-Jan Wagenmakers. 2012. 'Editors' Introduction to the Special Section on Replicability in Psychological Science a Crisis of Confidence?' *Perspectives on Psychological Science* 7 (6): 528–30.

Pescosolido, Bernice A., Tait R. Medina, Jack K. Martin, and J. Scott Long. 2013. 'The "Backbone" of Stigma: Identifying the Global Core of Public Prejudice Associated with Mental Illness'. *American Journal of Public Health* 103 (5): 853–60.

Petticrew, M., and H. Roberts. 2003. 'Evidence, Hierarchies, and Typologies: Horses for Courses'. *Journal of Epidemiology and Community Health* 57 (7): 527–9.

Pickering, Andrew. 2010. *The Cybernetic Brain: Sketches of Another Future*. Chicago, IL: University of Chicago Press.

Pies, Ronald W. 2014. 'The Bereavement Exclusion and DSM-5: An Update and Commentary'. *Innovations in Clinical Neuroscience* 11 (7–8): 19–22.

Piketty, Thomas. 2014. *Capital in the Twenty-First Century*. Cambridge, MA: Harvard University Press.

Ponthière, Grégory. 2003. 'Should We Discount Future Generations' Welfare? A Survey on the "Pure" Discount Rate Debate'. CREPP working papers 0302. HEC-Management School, University of Liège. Available from: https://ideas.repec.org/p/rpp/wpaper /0302.html.

Powers, Madison, and Ruth Faden. 2008. *Social Justice: The Moral Foundations of Public Health and Health Policy*. New York: Oxford University Press.

Pratt, Bridget, and Adnan A. Hyder. 2016. 'How Can Health Systems Research Reach the Worst-Off? A Conceptual Exploration'. *BMC Health Services Research* 16 (7): 619.

Preda, Adina. 2018. '"Justice in Health or Justice (and Health)?"—How (Not) to Apply a Theory of Justice to Health'. *Public Health Ethics* 11 (3): 336–45.

Privitera, Johanna. 2018. 'Aggregate Relevant Claims in Rescue Cases?' *Utilitas* 30 (2): 228–36.

Quammen, David. 2018. *The Tangled Tree: A Radical New History of Life*. London: William Collins.

Quinn, Warren. 1993. *Morality and Action*. Cambridge: Cambridge University Press.

Rachels, James. 1975. 'Active and Passive Euthanasia'. *The New England Journal of Medicine* 292 (2): 78–80.

Ramsey, Frank P. 1928. 'A Mathematical Theory of Saving'. *The Economic Journal* 38 (152): 543–59.

Rawls, John. 1993. *Political Liberalism*. New York: Columbia University Press.

Rawls, John. 1999. *A Theory of Justice*. Revised edition. Cambridge, MA: Harvard University Press.

Rawls, John. 2001. *Justice as Fairness: A Restatement*. Cambridge, MA: Harvard University Press.

Raz, Joseph. 1984. 'On the Nature of Rights'. *Mind* 93: 194–214.

Raz, Joseph. 1986. *The Morality of Freedom*. Oxford: Oxford University Press.

Reason, James. 1995. 'Understanding Adverse Events: Human Factors'. *BMJ Quality & Safety* 4 (2): 80–9.

Reibetanz, Sophia. 1998. 'Contractualism and Aggregation'. *Ethics* 108 (2): 296–311.

Reisig, V., and A. Hobbiss. 2000. 'Food Deserts and How to Tackle Them: A Study of One City's Approach'. *Health Education Journal* 59 (2): 137–49.

Review on Antimicrobial Resistance. 2014. *Antimicrobial Resistance: Tackling a Crisis for the Health and Wealth of Nations*. London: Her Majesty's Government. Available from: https://amr-review.org/Publications.html.

Review on Antimicrobial Resistance. 2016. *Tackling Drug-Resistant Infections Globally: Final Report and Recommendations*. London: Her Majesty's Government. Available from: https://amr-review.org/sites/default/files/160525_Final%20paper_with%20cover.pdf.

Rid, Thomas. 2016. *Rise of the Machines: A Cybernetic History*. New York: W. W. Norton & Company.

Roberts, Devender, Julie Brown, Nancy Medley, and Stuart R. Dalziel. 2017. 'Antenatal Corticosteroids for Accelerating Fetal Lung Maturation for Women at Risk of Preterm

Birth'. In Cochrane Database of Systematic Reviews, no. 3 (New York: John Wiley & Sons, Ltd). Available from: https://www.cochranelibrary.com/cdsr/doi/10.1002/14651858. CD004454.pub4/full?highlightAbstract=antenat%7Ccorticosteroid%7Cantenatal%7Ccorticosteroids.

Robeyns, Ingrid. 2017. *Wellbeing, Freedom and Social Justice: The Capability Approach Re-Examined*. Cambridge: Open Book Publishers. Available from: https://www.openbookpublishers.com/product/682.

Roemer, John E. 1993. 'A Pragmatic Theory of Responsibility for the Egalitarian Planner'. *Philosophy and Public Affairs* 22 (2): 146–66.

Roemer, John E., and Alain Trannoy. 2016. 'Equality of Opportunity: Theory and Measurement'. *Journal of Economic Literature* 54 (4): 1288–332.

Roosevelt, Franklin D. 1941. 'Executive Order 8807—Establishing the Office of Scientific Research and Development'. Available from: https://www.presidency.ucsb.edu/documents/executive-order-8807-establishing-the-office-scientific-research-and-development.

Rose, Geoffrey. 1981. 'Strategy of Prevention: Lessons from Cardiovascular Disease'. *BMJ* 282 (6279): 1847–51.

Rose, Geoffrey, and James McCormick. 2001. 'Sick Individuals and Sick Populations'. *International Journal of Epidemiology* 30 (3): 427–32.

Rosser, J. Barkley. 2015. 'Complexity and Austrian Economics'. In Christopher J. Coyne and Peter J. Boettke, eds, *The Oxford Handbook of Austrian Economics*, ch. 27. Oxford: Oxford University Press.

Rothwell, Peter. 2005. 'External Validity of Randomised Controlled Trials: "To Whom Do the Results of This Trial Apply?"'. *The Lancet* 365 (9453): 82–93.

Rudner, Richard. 1953. 'The Scientist Qua Scientist Makes Value Judgments'. *Philosophy of Science* 20 (1): 1–6.

Rumbold, Benedict. 2018. 'Towards a More Particularist View of Rights' Stringency'. *Res Publica* 25 (2): 211–33.

Rumbold, Benedict, Albert Weale, Annette Rid, James Wilson, and Peter Littlejohns. 2016. 'Public Reasoning and Health Care Priority Setting: The Case of NICE'. *Kennedy Institute of Ethics Journal* 27 (1): 107–34.

Rumbold, Benedict, and James Wilson. 2019. 'Privacy Rights and Public Information'. *Journal of Political Philosophy* 27 (1): 3–25.

Russo, Federica, and Jon Williamson. 2007. 'Interpreting Causality in the Health Sciences'. *International Studies in the Philosophy of Science* 21 (2): 157–70.

Rutter, Harry, Natalie Savona, Ketevan Glonti, Jo Bibby, Steven Cummins, Diane T. Finegood, Felix Greaves, et al. 2017. 'The Need for a Complex Systems Model of Evidence for Public Health'. *The Lancet* 390 (10112): 2602–4.

Sackett, David L., William M. C. Rosenberg, J. A. Muir Gray, R. Brian Haynes, and W. Scott Richardson. 1996. 'Evidence Based Medicine: What It Is and What It Isn't'. *BMJ* 312 (7023): 71–2.

Sackett, David L., Sharon E. Straus, W. Scott Richardson, William Rosenberg, and R. Brian Haynes. 2000. *Evidence-Based Medicine: How to Practice and Teach EBM*. 2nd edn. Edinburgh: Churchill Livingstone.

Scambler, Graham, and Anthony Hopkins. 1986. 'Being Epileptic: Coming to Terms with Stigma'. *Sociology of Health & Illness* 8 (1): 26–43.

Scanlon, T. M. 1998. *What We Owe to One Another*. Cambridge, MA: Harvard University Press.

Scanlon, T. M. 2018. *Why Does Inequality Matter?* Oxford: Oxford University Press.

Scheffler, Samuel. 2003. 'What Is Egalitarianism?' *Philosophy and Public Affairs* 31 (1): 5–39.

Schelling, Thomas C. 2006. *Micromotives and Macrobehavior*. New York: W. W. Norton.

Schwitzgebel, Eric, and Fiery Cushman. 2015. 'Philosophers' Biased Judgments Persist Despite Training, Expertise and Reflection'. *Cognition* 141: 127–37.

Segall, Shlomi. 2007. 'In Solidarity with the Imprudent: A Defense of Luck Egalitarianism'. *Social Theory and Practice* 33 (2): 177–98.

Segall, Shlomi. 2009. *Health, Luck, and Justice*. Princeton, NJ: Princeton University Press.

Selgelid, Michael J. 2005. 'Ethics and Infectious Disease'. *Bioethics* 19 (3): 272–89.

Sen, Amartya. 1979. 'Equality of What? The Tanner Lectures on Human Values'. Available from: https://tannerlectures.utah.edu/_documents/a-to-z/s/sen80.pdf.

Sen, Amartya. 1999. *Development as Freedom*. Oxford: Oxford University Press.

Sen, Amartya. 2004. 'Why Health Equity?' In Sudhir Anand, Fabienne Peter, and Amartya Sen, eds, *Public Health, Ethics, and Equity*, 21–34. Oxford: Oxford University Press.

Sen, Amartya. 2006. 'What Do We Want from a Theory of Justice?' *The Journal of Philosophy* 103 (5): 215–38.

Sen, Amartya. 2011. *The Idea of Justice*. Cambridge, MA: Harvard University Press.

Shafer-Landau, Russ. 2005. 'Liberalism and Paternalism'. *Legal Theory* 11 (3): 169–91.

Shavers, V. L. 2007. 'Measurement of Socioeconomic Status in Health Disparities Research'. *Journal of the National Medical Association* 99 (9): 1013–23.

Sheehan, Mark. 2007. 'Resources and the Rule of Rescue'. *Journal of Applied Philosophy* 24 (4): 352–66.

Shiffrin, Seana Valentine. 2000. 'Paternalism, Unconscionability Doctrine, and Accommodation'. *Philosophy and Public Affairs* 29 (3): 205–50.

Shue, Henry. 1996. *Basic Rights: Subsistence, Affluence, and US Foreign Policy*. 2nd edn. Princeton, NJ: Princeton University Press.

Simester, Andrew P., ed. 2005. *Appraising Strict Liability*. Oxford: Oxford University Press.

Simon, Herbert A. 1962. 'The Architecture of Complexity'. *Proceedings of the American Philosophical Society* 106 (6): 467–82.

Singer, Peter. 1972. 'Famine, Affluence, and Morality'. *Philosophy and Public Affairs* 1 (3): 229–43.

Sinnott-Armstrong, W. 2005. 'It's Not *My* Fault: Global Warming and Individual Moral Obligations'. In Walter Sinnott-Armstrong and Richard B. Howarth, eds, *Perspectives on Climate Change: Science, Economics, Politics, Ethics (Advances in the Economics of Environmental Resources, Vol. 5)*, 285–307. Bingley: Emerald Group Publishing Limited.

Sloan Wilson, David. 2016. 'Two Meanings of Complex Adaptive Systems'. In David Sloan Wilson and Alan Kirman, eds, *Complexity and Evolution: Toward a New Synthesis for Economics*, 31–46. Cambridge, MA: MIT Press.

Sloan Wilson, David, and John M. Gowdy. 2015. 'Human Ultrasociality and the Invisible Hand: Foundational Developments in Evolutionary Science Alter a Foundational Concept in Economics'. *Journal of Bioeconomics* 17 (1): 37–52.

Smart, J. J. C., and Bernard Williams. 1973. *Utilitarianism: For and Against*. Cambridge: Cambridge University Press.

Smith, Adam. 1982 [1790]. *The Theory of Moral Sentiments*, edited by D. D. Raphael and A. L. Macfie, vol. I of the Glasgow Edition of the Works and Correspondence of Adam Smith. Carmel, IN: Liberty Fund.

Smith, Gordon C. S., and Jill P. Pell. 2003. 'Parachute Use to Prevent Death and Major Trauma Related to Gravitational Challenge: Systematic Review of Randomised Controlled Trials'. *BMJ* 327 (7429): 1459–61.

Smith, Maxwell J. 2015. 'Health Equity in Public Health: Clarifying Our Commitment'. *Public Health Ethics* 8 (2): 173–84.

Snowdon, Christopher. 2017. *Killjoys: A Critique of Paternalism*. London: The Institute of Economic Affairs.

Sommer, Marni, and Richard Parker, eds. 2013. *Structural Approaches in Public Health*. Abingdon: Taylor & Francis.

Sousa, Michael D. 2018. 'The Persistence of Bankruptcy Stigma'. *American Bankruptcy Institute Law Review* 26: 217–42.

Sreenivasan, Gopal. 2012. 'A Human Right to Health? Some Inconclusive Scepticism'. *Aristotelian Society Supplementary* 86 (1): 239–65.

Stanger-Hall, Kathrin F., and David W. Hall. 2011. 'Abstinence-Only Education and Teen Pregnancy Rates: Why We Need Comprehensive Sex Education in the U.S'. *PLoS ONE* 6 (10): e24658.

Steiner, Michael. 2020. 'Coronavirus Has Changed How We Support People with Failing Lungs – a Doctor Explains Why'. *The Conversation*. Available from: http://theconversation.com/coronavirus-has-changed-how-we-support-people-with-failing-lungs-a-doctor-explains-why-149960.

Stepan, Nancy Leys. 2011. *Eradication: Ridding the World of Diseases Forever?* London: Reaktion Books.

Sterman, John D. 2001. 'System Dynamics Modeling: Tools for Learning in a Complex World'. *California Management Review* 43 (4): 8–25.

Sterman, John D. 2006. 'Learning from Evidence in a Complex World'. *American Journal of Public Health* 96 (3): 505–14.

Stilgoe, Jack. 2014. 'Against Excellence'. *The Guardian*, 19 December. Available from: https://www.theguardian.com/science/political-science/2014/dec/19/against-excellence.

Stodden, Victoria. 2014. 'Enabling Reproducibility in Big Data Research: Balancing Confidentiality and Scientific Transparency'. In Julia Lane, Victoria Stodden, Stefan Bender, and Helen Nissenbaum, eds, *Privacy, Big Data and the Public Good*, 112–32. Cambridge: Cambridge University Press.

Stokes, Donald E. 1997. *Pasteur's Quadrant: Basic Science and Technological Innovation*. Washington, DC: Brookings Institution.

Stoye, George. 2018. 'The NHS at 70: Does the NHS Need More Money and How Could We Pay for It?' The King's Fund. Available from: https://www.kingsfund.org.uk/publications/nhs-70-does-the-nhs-need-more-money.

SAGE [Strategic Advisory Group of Experts on Immunization]. 2014. *Report of the SAGE Working Group on Vaccine Hesitancy*. Geneva: World Health Organization. Available from: https://www.who.int/immunization/sage/meetings/2014/october/SAGE_working_group_revised_report_vaccine_hesitancy.pdf?ua=1.

Strawson, P. F. 1974. *Freedom and Resentment and Other Essays*. London: Routledge.

Stutzin Donoso, Francisca. 2018. 'Chronic Disease as Risk Multiplier for Disadvantage'. *Journal of Medical Ethics* 44 (6): 371–5.

Syvanen, Michael. 2012. 'Evolutionary Implications of Horizontal Gene Transfer'. *Annual Review of Genetics* 46: 341–58.

Taurek, John M. 1977. 'Should the Numbers Count?' *Philosophy and Public Affairs* 6 (4): 293–331.

Taylor, Mark J., and James Wilson. 2019. 'Reasonable Expectations of Privacy and Disclosure of Health Data'. *Medical Law Review* 27 (3): 432–60.

Temkin, Larry S. 1993. *Inequality*. New York: Oxford University Press.

Thaler, Richard H., and Cass R. Sunstein. 2008. *Nudge: Improving Decisions about Health, Wealth, and Happiness.* New Haven, CT: Yale University Press.

Thompson, Christopher. 2018. 'Rose's Prevention Paradox'. *Journal of Applied Philosophy* 35 (2): 242–56.

Thompson, Kimberly M., and Radboud J. Duintjer Tebbens. 2007. 'Eradication Versus Control for Poliomyelitis: An Economic Analysis'. *Lancet* 369 (9570): 1363–71.

Thomson, Judith Jarvis. 1971. 'A Defense of Abortion'. *Philosophy and Public Affairs* 1 (1): 47–66.

Thomson, Judith Jarvis. 1976. 'Killing, Letting Die, and the Trolley Problem'. *The Monist* 59 (2): 204–17.

Tomasello, Michaelz. 2016. *A Natural History of Human Morality.* Cambridge: MA: Harvard University Press.

Tomlin, Patrick. 2017. 'On Limited Aggregation'. *Philosophy and Public Affairs* 45 (3): 232–60.

Tudor-Hart, Julian. 1971. 'The Inverse Care Law'. *The Lancet* 297 (7696): 405–12.

Ulijaszek, Stanley J. 2017. *Models of Obesity: From Ecology to Complexity in Science and Policy.* Cambridge: Cambridge University Press.

UN Committee on Economic Social and Cultural Rights. 2000. 'General Comment No. 14 (2000), The Right to the Highest Attainable Standard of Health (Article 12 of the International Covenant on Economic, Social and Cultural Rights)'. New York: United Nations. Available from: https://digitallibrary.un.org/record/425041.

Unger, Peter. 1996. *Living High and Letting Die: Our Illusion of Innocence.* Oxford: Oxford University Press.

United Nations Centre for Human Rights. 1989. *Right to Adequate Food as a Human Right.* E/CN.4/Sub.2/1987/23. New York: United Nations.

Varoufakis, Yanis. 2016. 'Transcript: Interview with Yanis Varoufakis'. *The Economist.* Availableat:https://www.economist.com/briefing/2016/03/31/transcript-interview-with-yanis-varoufakis.

Venkatapuram, Sridhar. 2011. *Health Justice: An Argument from the Capabilities Approach.* Cambridge: Polity Press.

Verweij, Marcel, and Angus Dawson. 2004. 'Ethical Principles for Collective Immunisation Programmes'. *Vaccine* 22 (23): 3122–6.

Verweij, Marcel, and Angus Dawson. 2009. 'The Meaning of "Public" in "Public Health"'. In Angus Dawson and Marcel Verweij, eds, *Ethics, Prevention, and Public Health*, 13–29. Oxford: Oxford University Press.

Vogt, Henrik, Sara Green, Claus Thorn Ekstrøm, and John Brodersen. 2019. 'How Precision Medicine and Screening with Big Data Could Increase Overdiagnosis'. *BMJ* 366 (September): l5270.

Voigt, Kristin. 2013. 'Appeals to Individual Responsibility for Health'. *Cambridge Quarterly of Healthcare Ethics* 22 (2): 146–58.

von Neumann, John. 1966. *Theory of Self-Reproducing Automata.* Edited by Arthur W. Burks. Urbana: University of Illinois Press.

Voorhoeve, Alex. 2009. *Conversations on Ethics.* Oxford: Oxford University Press.

Voorhoeve, Alex. 2014. 'How Should We Aggregate Competing Claims?' *Ethics* 125 (1): 64–87.

Voorhoeve, Alex, and Marc Fleurbaey. 2012. 'Egalitarianism and the Separateness of Persons'. *Utilitas* 24 (3): 381–98.

Wainer, Howard. 2009. *Picturing the Uncertain World: How to Understand, Communicate, and Control Uncertainty Through Graphical Display.* Princeton, NJ: Princeton University Press.

Waldron, Jeremy. 1992. 'From Authors to Copiers: Individual Rights and Social Values in Intellectual Property'. *Chicago Kent Law Review* 68: 841–88.

Waldron, Jeremy. 2017. *One Another's Equals: The Basis of Human Equality*. Cambridge, MA: Harvard University Press.

Waldrop, M. Mitchell. 1992. *Complexity: The Emerging Science at the Edge of Order and Chaos*. New York: Simon & Schuster.

Walker, Tom. 2019. *Ethics and Chronic Illness*. London: Routledge.

Wang, Haidong, Amanuel Alemu Abajobir, Kalkidan Hassen Abate, Cristiana Abbafati, Kaja M. Abbas, Foad Abd-Allah, Semaw Ferede Abera, et al. 2017. 'Global, Regional, and National Under-5 Mortality, Adult Mortality, Age-Specific Mortality, and Life Expectancy, 1970–2016: A Systematic Analysis for the Global Burden of Disease Study 2016'. *The Lancet* 390 (10100): 1084–150.

Weatherford, Jack. 1997. *History of Money: From Sandstone to Cyberspace*. New York: Crown Publishers.

Weaver, Warren. 1948. 'Science and Complexity'. *American Scientist* 36 (4): 536–44.

White, Douglas B., and Bernard Lo. 2020. 'A Framework for Rationing Ventilators and Critical Care Beds During the COVID-19 Pandemic'. *JAMA* 323 (18): 1773–4.

Whitehead, Margaret. 1990. *The Concepts and Principles of Equity and Health*. Copenhagen: WHO Regional Office for Europe. Document number: EUR/ICP/RPD 414.

Whitehead, Margaret, and Göran Dahlgren. 2006. *Concepts and Principles for Tackling Social Inequities in Health: Levelling Up Part 1*. Copenhagen: WHO Regional Office for Europe. Available from: https://www.euro.who.int/__data/assets/pdf_file/0010/74737/E89383.pdf.

Wigglesworth, Robin. 2018. 'How a Volatility Virus Infected Wall Street'. *Financial Times*, 12 April. Available from: https://www.ft.com/content/be68aac6-3d13-11e8-b9f9-de94fa33a81e.

Wild, Verina, D. Jaff, N. S. Shah, and M. Frick. 2017. 'Tuberculosis, Human Rights and Ethics Considerations Along the Route of a Highly Vulnerable Migrant from Sub-Saharan Africa to Europe'. *International Journal of Tuberculosis and Lung Disease* 21 (10): 1075–85.

Wilkinson, Martin, and Andrew Moore. 1997. 'Inducement in Research'. *Bioethics* 11 (5): 373–89.

Wilkinson, Martin, and Andrew Moore. 1999. 'Inducements Revisited'. *Bioethics* 13 (2): 114–30.

Wilkinson, Richard, and Michael Marmot. 2003. *Social Determinants of Health: The Solid Facts*. 2nd edn. New York: World Health Organization. Available from: http://www.euro.who.int/__data/assets/pdf_file/0005/98438/e81384.pdf.

Williams, Bernard. 1981. *Moral Luck: Philosophical Papers 1973–1980*. Cambridge: Cambridge University Press.

Williams, Garrath. 2008. 'Responsibility as a Virtue'. *Ethical Theory and Moral Practice* 11 (4): 455–70.

Wilson, James. 2007a. 'Is Respect for Autonomy Defensible?' *Journal of Medical Ethics* 33 (6): 353–6.

Wilson, James. 2007b. 'Rights'. In Richard E. Ashcroft, Angus Dawson, Heather Draper, and John McMillan, eds, *Principles of Health Care Ethics*, 2nd edn, 239–46. Chester: John Wiley and Sons.

Wilson, James. 2007c. 'Nietzsche and Equality'. In Gudrun von Tevenar, ed., *Nietzsche and Ethics*, 221–40. Oxford: Peter Lang.

Wilson, James. 2007d. 'Transhumanism and Moral Equality'. *Bioethics* 21 (8): 419–25.

Wilson, James. 2009a. 'Could There Be a Right to Own Intellectual Property?' *Law and Philosophy* 28: 393–427.

Wilson, James. 2009b. 'Not so Special After All? Daniels and the Social Determinants of Health'. *Journal of Medical Ethics* 35 (1): 3–6.

Wilson, James. 2011a. 'Why It's Time to Stop Worrying about Paternalism in Health Policy'. *Public Health Ethics* 4 (3): 269–79.

Wilson, James. 2011b. 'Health Inequities'. In Angus Dawson, ed., *Public Health Ethics: Key Concepts and Issues in Policy and Practice*, 211–30. Cambridge: Cambridge University Press.

Wilson, James. 2012. 'Paying for Patented Drugs Is Hard to Justify: An Argument about Time Discounting and Medical Need'. *Journal of Applied Philosophy* 29 (3): 186–99.

Wilson, James. 2013. 'Drug Resistance, Patents and Justice: Who Owns the Effectiveness of Antibiotics?' In John Coggon and Swati Gola, eds, *Global Health and International Community: Ethical, Political and Regulatory Challenges*, ch. 9. London: Bloomsbury Academic.

Wilson, James. 2014. 'Embracing Complexity: Theory, Cases and the Future of Bioethics'. *Monash Bioethics Review* 32 (1–2): 3–21.

Wilson, James. 2016. 'Public Value, Maximization and Health Policy: An Examination of Hausman's Restricted Consequentialism'. *Public Health Ethics* 10 (2): 157–63.

Wilson, James. 2018. 'Global Justice'. In Dominick A. Dellasala and Michael I. Goldstein, eds, *Encyclopedia of the Anthropocene*, 81–6. Oxford: Elsevier.

Wilson, James. 2020. 'Philanthrocapitalism and Global Health'. In Gillian Brock and Solomon Benatar, eds, *Global Health: Ethical Challenges*, 416–28. Cambridge: Cambridge University Press.

Wilson, James. 2021. 'When Does Precision Matter? Personalised Medicine from the Perspective of Public Health'. In Margherita Brusa, Michael Barilan, and Aaron Ciechanover, eds, *Can Precision Medicine Be Personal; Can Personalized Medicine Be Precise?* Oxford: Oxford University Press.

Wilson, James, and David Hunter. 2010. 'Research Exceptionalism'. *American Journal of Bioethics* 10 (8): 45–54.

Wilson, James, and David Hunter. 2011. 'Hyper-Expensive New Therapies and the Prioritisation of R&D'. London: Nuffield Council on Bioethics. Available from: https://discovery.ucl.ac.uk/id/eprint/1325654/.

Winslow, C.-E. A. 1903. 'Statistics of Small-Pox and Vaccination'. *Publications of the American Statistical Association* 8 (61): 279–84.

Wolfe, Robert M, and Lisa K Sharp. 2002. 'Anti-Vaccinationists Past and Present'. *BMJ* 325 (7361): 430.

Wolff, Jonathan. 1998. 'Fairness, Respect, and the Egalitarian Ethos'. *Philosophy and Public Affairs* 27 (2): 97–122.

Wolff, Jonathan. 2015. 'Social Equality and Social Inequality'. In Carina Fourie, Fabian Schuppert, and Ivo Wallimann-Helmer, eds, *Social Equality: On What It Means to Be Equals*, 209–25. Oxford: Oxford University Press.

Wolff, Jonathan, and Avner De-Shalit. 2007. *Disadvantage*. Oxford: Oxford University Press.

Wolff, Robert Paul. 1970. *In Defense of Anarchism*. Berkeley: University of California Press.

Wootton, David. 2007. *Bad Medicine: Doctors Doing Harm Since Hippocrates*. Oxford: Oxford University Press.

World Bank. 2019a. 'Current Health Expenditure (% of GDP)'. Available from: https://data.worldbank.org/indicator/SH.XPD.CHEX.GD.ZS.

World Bank. 2019b. 'Life Expectancy at Birth, Total (Years)'. Available from: https://data. worldbank.org/indicator/SP.DYN.LE00.IN?end=2017&locations=CU-US-CR&start= 1960&view=chart.

World Health Organization. 1948. 'Preamble to the Constitution of the World Health Organization as Adopted by the International Health Conference, New York, 19–22 June, 1946'. New York: World Health Organization. Available from: https://www.who. int/governance/eb/who_constitution_en.pdf.

World Health Organization. 2011. *Scaling Up Action Against Noncommunicable Diseases: How Much Will It Cost?* Geneva: World Health Organization.

World Health Organization. 2013. *WHO Global Status Report on Road Safety 2013: Supporting a Decade of Action.* World Health Organization: New York. Available from: http://www.who.int/iris/bitstream/10665/78256/1/9789241564564_eng.pdf.

World Health Organization. 2018a. *World Malaria Report 2018.* Geneva: World Health Organization. Available from: https://www.who.int/malaria/publications/world-malaria-report-2018/en.

World Health Organization. 2018b. 'Antimicrobial Resistance: Fact Sheet'. Available from: https://www.who.int/news-room/fact-sheets/detail/antimicrobial-resistance.

World Health Organization. 2019a. 'Ten Threats to Global Health in 2019'. Available from: https://www.who.int/news-room/spotlight/ten-threats-to-global-health-in-2019.

World Health Organization. 2019b. 'Fact Sheet: Measles'. Available from: https://www.who. int/en/news-room/fact-sheets/detail/measles.

World Medical Association. 2006. 'World Medical Association International Code of Medical Ethics'. Available from: https://www.wma.net/policies-post/wma-international-code-of-medical-ethics/.

World Obesity Federation. 2020. 'Global Obesity Observatory'. Available from: https:// www.worldobesitydata.org.

Wu, Joseph. 2020. 'The Limits of Screening'. PhD Thesis, University of Cambridge. Available from: https://www.repository.cam.ac.uk/handle/1810/306007.

Yong, Ed. 2016. *I Contain Multitudes: The Microbes Within Us and a Grander View of Life.* New York: Random House.

Young, Iris Marion. 2001. 'Equality of Whom? Social Groups and Judgments of Injustice'. *Journal of Political Philosophy* 9 (1): 1–18.

Young, Iris Marion. 2013. *Responsibility for Justice.* Oxford: Oxford University Press.

Zachar, Peter, and Kenneth S. Kendler. 2017. 'The Philosophy of Nosology'. *Annual Reviews of Clinical Psychology* 13: 49–71.

Zuckerman, Harriet A. 1965. 'Nobel Laureates in the United States: A Sociological Study of Scientific Collaboration'. Unpublished PhD dissertation, Columbia University.

Index

For the benefit of digital users, indexed terms that span two pages (e.g., 52–53) may, on occasion, appear on only one of those pages.

Printed and bound by CPI Group (UK) Ltd, Croydon, CR0 4YY